METHODS AND STYLES IN THE DEVELOPMENT OF CHEMISTRY

METHODS
AND
STYLES
IN THE
DEVELOPMENT
OF
CHEMISTRY

Joseph S. Fruton

American Philosophical Society
Philadelphia • *2002*

Memoirs
of the
American Philosophical Society
Held at Philadelphia
For Promoting Useful Knowledge
Volume 245

Library of Congress Cataloging-in-Publication Data

Fruton, Joseph S (Joseph Stewart), 1912-
    Methods and styles in the development of chemistry /
Joseph S. Fruton.
        p.   cm. — (Memoirs of the American Philosophical
Society held at Philadelphia for promoting useful knowledge,
ISSN 0065-9738 ; v. 245)
Includes bibliographical references and index.
ISBN 0-87169-245-7 (cloth)
    1. Chemistry—History. I. Title. II. Memoirs of the American
Philosophical Society ; v. 245.

Q11 .P612 vol. 245
[QD11]
540'.9—dc21
                                                    2001045791

I am greatly indebted to
Otto Theodore Benfey, Jerome A. Berson, Roald Hoffmann,
Frederic L. Holmes, and Alan J. Rocke
for their valuable criticisms of and suggestions
for an early draft of this book.

CONTENTS

ILLUSTRATIONS

FOREWORD

Lack of experience diminishes our power of taking a comprehensive view of the admitted facts. Hence those who dwell in intimate association with nature and its phenomena grow more and more able to formulate, as the foundation of their theories, principles such as to admit of a wide and coherent development; while those whom devotion to abstract discussions has rendered unobservant of the facts are too ready to dogmatize on the basis of a few observations. [1]

IN THIS BOOK, I consider the varieties of methods and styles in the thought and practice of some of the many people who investigated the general properties of matter, the specific properties of material objects (including living organisms), and the phenomena exhibited when such objects are subjected to artificial treatment. Over the centuries, the efforts of these people, belonging to different professions—physicians, natural philosophers, alchemists, pharmacists, metallurgists, mineralogists, and in modern times, chemists, physicists, and biologists of various kinds—contributed to the development of the field of scientific inquiry we now call "chemistry."

The word "style" has come to have many different meanings in various contexts—haute couture, the literary, visual, and musical arts, as well as politics and other social activities. It has also been used as a euphemism for personal idiosyncracy.[2] I use "style" as a neutral term denoting the manner in which chemical investigation was conducted. Its meaning is not restricted to the aesthetic attributes (simplicity, symmetry) of particular chemical theories, to their "truth" or "falsity," or to the clarity of their exposition. Nor is "style" applied solely to chemical practice in which exceptional originality, skill, ingenuity, or precision was exhibited in the use and development of experimental methods. Such elegance was rightly hailed by contemporaries and by later historians, but imaginative (often illogical) thought, as well as routine or imprecise laboratory work, have also played a significant role in the historical development of chemical knowledge.

In the conduct of their chemical studies, and in the interpretation of the results, the investigators were influenced by preconceptions derived from many sources, among them religious belief, philosophical and technical tradition, previous education and practical experience, and social custom. As Stephen Toulmin put it:

> However objective and open-minded a scientist may be, his questions necessarily involve the use of preformed concepts. In this sense, he relies on preconceived ideas; but it is a sense which emphatically does not involve any suggestion of dogmatism. For these concepts are the common coin of scientific discussion in a given period, the basis of common understanding between scientists and of their language; they are essential if there is to be any agreement between scientists about what questions are relevant, sensible, even intelligible. [3]

In what follows, I consider the thought and action of these people, whether or not they are now considered to have been "scientists," to form a continuum in the development of chemical knowledge, leading to the multiplicity of specialties and sub-specialties that now constitute the chemical sciences. In connection with the changes in the conceptual structure of an area of chemical inquiry, I emphasize the practice that led to the discovery of previously unknown natural objects and phenomena, to the performance of fruitful experiments, to the invention of new or better methods, tools, and instruments, and to changes in classification, nomenclature and symbolism. I also consider, where appropriate, biographical data relevant to the theme of this essay. In denoting the preconceptions of an individual, I do not use the terms "naive realist," "empiricist," "positivist," "presentist," "pragmatist," "whig," "conventionalist," etc., which are frequently used (and misused) by historians. All too often, what they have taken to be an intellectual commitment may have been, for the scientist, a "working hypothesis." Nor do I use the currently fashionable (but frequently misapplied) phrases "material culture" or "moral economy." Scientist-historians such as I have been chastised by some professional historians of science as being indifferent to archival sources, especially personal correspondence, and I agree that the critical study of such material is essential in historical research, as will be evident in later pages of this book. I must add, however, that it is by no means certain in all cases that the selected correspondence of a noted scientist reveals more about his or her method and style of thought and work than a well-informed and thorough examination of the entire published scientific output. [4]

Distinctions have been made between the styles of scientific thought in German-speaking, French-speaking, and English-speaking nations. [5] Also, much attention has been paid to the idea of a "Denkstil," offered by the serologist Ludwik Fleck, and derived from the writings of the sociologist Karl Mannheim. [6] The most influential recent theory is the one presented by Thomas Kuhn, who set forth stimulating views about what he called "normal science," characterized by "puzzle-solving" along lines set by prevailing concepts ("paradigms") in a scientific community, and the appearance of "anomalies" whose explanation led to a revolution and to the adoption of new paradigms.[7] The recent discussion of the eighteenth-century Chemical Revolution (of which more later) has raised questions about the applicability of Kuhn's theory to the historical development of the chemical sciences. Moreover, in relation to the theme of this book, the concept of a scientific community, in which particular paradigms prevail, glosses over the differences in the methods and styles of the individual members of that community.[8] I was reminded of the warning by Pieter Geyl that a modern historian should be

> on his guard against . . . intellectually satisfying schemes which may hedge in or distort the view. . . . He will not too readily identify a period with an idea; behind the idea he will look for the unruly, struggling men. Behind the anonymity of a class, of a nation, of a sect he will search for various shadings, for individual peculiarities.[9]

For my purpose, a more valuable guide has been the recent magisterial three-volume treatise by the late historian Alistair Crombie,[10] in which he traced the emergence in Europe of six styles of scientific reasoning: mathematical, experimental, analogical modeling, taxonomic, probabilistic and statistical, and historical:

> These six styles and their objects are all different, sometimes incommensurable, assuming fundamentally different worlds, but frequently they are combined in any particular research. By identifying the regularities that become its object of inquiry, and by defining its questions and acceptable evidence and answers, a style both creates its own subject-matter and is created by it. A change of style introduces not only new subject-matter, but also new questions about the same subject-matter. . . . Different styles introduce new questions about the existence of their theoretical objects; are these real or products of methods of measurement or sampling, or even of language?[11]

In his treatise, Crombie discusses in considerable detail examples drawn largely from the physics and physiology developed by European investigators before the end of the nineteenth century. As will be evident in what follows, each of the six styles is represented in the historical development of chemical knowledge, but the rapidity and extent of the transformation of chemistry after 1800 led many investigators to change the style of their research. For this reason, I deem it necessary to retell some of what is well known to professional historians of chemistry, in order to provide the context in which the individual styles were expressed. Also, since I will consider the styles of several chemists active during the twentieth century, when the award of Nobel Prizes has often affected historical judgment, some of the relevant theoretical and practical background will also need retelling.

To the styles of thinking listed by Crombie might be added some less well-defined mental attributes, such as imagination or intuition in the choice of chemical problems, of experimental attack and interpretation of experimental results, or of nomenclature and symbolism.[12] Among the styles related to chemical practice, one might include features such as technical skill, ingenuity, precision, or even dependence on group effort in the laboratory. Each chemical style has had its own standards in the evaluation of theories and of experimental data. With the development of a research field, these standards have changed, although vestiges of older styles, such as Pythagorean numerology or the laboratory operations of medieval alchemists, may have been retained.

This book deals, therefore, primarily with the methods and styles of individual scientists in the formulation of their theories, in the conduct of the experimental work in their laboratories, in the presentation of the results of their thought and action, as well as in their acknowledgment of prior work on the problem at hand, and in their response to contemporary critics. Clearly, the differences in style depended to some degree on the extent and quality of their early education, which was related to the social status of their parents or guardians. The men from wealthy families went to leading universities, or had competent private tutors, and often were provided with private laboratories. The less fortunate ones, who emerged from poverty, were largely self-educated, learned laboratory methods as apprentices, and likely were sustained by religious belief. The majority came from middle-class families, and frequently tended to be more competitive than those in the two other groups.

To the above introductory remarks, I should add that professional historians of the natural sciences have differed in their preconceptions of what constitutes "good history."[13] In recent decades there has been argument about such matters as "internalism" versus "externalism," "presentism," "instrumentalism," the "Whig interpretation of history," and the "social construction of scientific knowledge."[14] The debate has now lessened somewhat in intensity, with the acceptance of the view that valuable insight into the complexity and special nature of the development of the natural sciences may be gained from different approaches—philosophical, technical, sociological, and biographical. What still divides professional historians from scientists who study the historical development of their specialty is the tendency of many of the former to overgeneralize their interpretation of selected historical data, and the tendency of the latter to emphasize the theoretical and empirical aspects of particular scientific problems, without consideration of their "cultural" significance.

It has recently become fashionable among philosophers and historians to write about the "disunity of science" and to call attention to the differences in thought and practice among the areas of scientific investigation now denoted as separate "disciplines."[15] Such differences were recognized by Auguste Comte and William Whewell, but the attraction of the idea of the "unity of science" (implicit in the natural philosophy of Aristotle) became pervasive at the end of the nineteenth century. Among its proponents were the scientists Hermann Helmholtz and Ernst Mach (both physicists and physiologists), and their views inspired the so-called Vienna Circle of "logical positivists."

One consequence of the adoption of the idea of the unity of science has been the present-day institutional classification of chemistry as a "physical science." During most of the nineteenth century, however, the dominant chemical specialty was "organic" chemistry, defined by Jöns Jacob Berzelius as dealing with the study of "organized" matter, with close links to animal and plant anatomy and physiology.[16] This tie to what came to be called "biology" (now often termed "life sciences") was also evident in the various schemes for the classification of the innumerable known chemical substances, both "organic" and "inorganic." The relationship of chemistry to physics was expressed during the seventeenth and eighteenth centuries and the first half of the nineteenth century in the term "philosophical chemistry," which included experimental study of heat and electricity.[17] The emergence after about 1850 of the mathematical treatment of "energy" (thermodynamics) and of electromagnetism led to the establishment of "physical chemistry" (or "theoretical chemistry") as a distinctive institutional discipline,[18] the links to experimental physics having been established earlier in the century, most notably by Michael Faraday. During the twentieth century, with the application of quantum theory and wave mechanics to problems of chemical structure, some physicists preferred the term "chemical physics." In similar fashion, later in the century, many biologists preferred "molecular biology" to "biochemistry."[19]

Implicit in the idea of the "unity of science" is the question of whether chemistry has been (or can be) "reduced" to physics, or whether biology has been (or can be) "reduced" to chemistry.[20] The philosophers and philosophically-minded scientists who have argued about this question have largely focused attention on the theoretical physics or physical chemistry of their time, or on biological problems arising from the study of evolution or genetics. Although they may have recognized that chemistry — the study of the specific properties ("forms and qualities") and transformations of the millions of substances found in nature and those made by man — fits into a continuum of scientific knowledge that ranges from the general properties of matter to the specific properties of living organisms, relatively few modern philosophers have ventured to examine the philosophical aspects of the development of chemical thought and practice, and appear to have followed Kant in denying to chemistry the status of a "science."[21]

Another recent fashion has been the "laboratory study" of scientific practice, especially the nature of experiment, as a reaction to the past tendency to emphasize the development of scientific thought.[22] Most of the "case studies" were drawn from the work of physicists and biologists, and only occasionally from that of chemists. This new interest in scientific practice was in part an outgrowth of the attempt of some sociologists to show that scientific knowledge is "socially constructed."[23] The members of this group disparaged the important studies of Robert Merton and others on the social customs and conduct of scientists.[24]

Chapter One
THE GREEK INHERITANCE AND ALCHEMY

ᴐ

AMONG THE STYLES of mathematical reasoning about natural objects and phenomena, the oldest appears to be the numerology that came from ancient crafts—metal-working, gold-refining, manufacture of pottery or glass, dyeing, mummification—or the use of eyes and hands to measure and compare quantities such as the number of days or of flocks of domestic animals, and the use of stones of different size to determine the relative weight of objects.[1] By 500 B.C. such numerology had been highly developed in the Middle East and Asia (Egypt, Babylonia, Persia, India) where apart from their utility in everyday life, numbers acquired a religious character linked to astronomical observation, as in the identification of the number of "moving stars" in the heavens with that of the membership of a particular pantheon. Certain numbers, such as 1 (for the monotheists), 4, or 7, and their combinations, were assigned special theological importance. According to ancient tradition, in about 500 B.C. Pythagoras of Samos founded a secret cult whose members developed his theory of numbers, based mainly on musical harmonies and the movements of heavenly bodies.[2] What has been termed the "hypnotic power of numerology"[3] has been a recurrent feature of chemical speculation during the succeeding centuries, and I will mention later examples that have appeared in the writings of prominent scientists.

Their aim may have been solely to seek simplicity and generality, but some of them, in writing for the public, echoed Plato's Timaeus in statements such as astronomer James Jeans': "The Great Architect of the Universe now begins to appear as a pure mathematician."[4]

Although it has long been customary to refer to the *use* of the theory of numbers as "applied mathematics," the origins of many other important mathematical ideas can also be traced to ancient practical human activities.[5] The centuries-old observation of the shapes of terrestrial and celestial objects, and everyday tasks such as land measurement, also produced geometry, as expounded in Euclid's *Elements*, still a model of a book on "pure" mathematics. The Greek style of mathematical thought was carried forward by the Roman philosopher Boethius, who was credited in medieval translations of Greek and Arabic texts with having defined the *quadrivium* of mathematical studies—arithmetic, music, geometry, astronomy. This definition was developed by European scholars, notably Nicholas of Cusa, into a Pythagorean-Platonic-Christian approach to the study of the way God created the world.[6] Many geometric figures appeared in the symbolism of medieval alchemists[7] and in the philosophical writings of Descartes, but it was not until the nineteenth century that geometric concepts became a prominent feature of thought about the structure of chemical molecules. Likewise, the calculus, which arose from the empirical study of motion and change, and involved the new concepts of variable magnitudes and functions (Galileo, Descartes, Newton, Leibniz), also did not figure largely in chemical theory until after about 1850. Indeed, in 1838 Auguste Comte wrote: "Every attempt to refer chemical questions to mathematical doctrines must be considered, now as always, profoundly irrational, as being contrary to the nature of the phenomena. . . . It would occasion vast and rapid retrogradation, by substituting vague conceptions for positive ideas, and an easy algebraic verbiage for a laborious investigation of facts." [8]

The origins of the experimental style of chemical reasoning also lie in the centuries-long activities of prehistoric craftsmen and physicians. During the centuries before Plato and Aristotle, Greek philosophers (notably Thales, Anaximander, Anaximenes, Empedocles, Leucippus, and Democritos) based their theories about the fundamental units of matter on the practical knowledge that had been gained, through trial and error, in efforts to improve the quality of products—as in the refining of gold—or the treatment of disease. These pre-Socratic philosophers derived from this empirical knowledge concepts represented by the present-day English words "substance," "element," "principle," and "atom." [9] In the sixth century B.C. Thales assumed that the basic principle is water, Anaximander that it is *apeiron* (moist air?), and Anaximenes that it is *pneuma* (breath). About a hundred years later, Empedocles advanced the idea (which lasted well into the eighteenth century A.D.) that all things have four "roots"—fire and air (which rose upward), water and earth (which fell downward). These "elements" were associated with the "active qualities" hot and cold, and the "passive qualities" wet and dry, and in medical practice with the four "humors" yellow bile and blood, phlegm, and black bile. Leucippus and Democritus defined atoms as hard, indivisible particles of variable size moving in empty space. In his *Timaeus*, Plato used Pythagorean numerology and geometry to argue that the fundamental entities were not units of matter but ideal Forms created by God. [10] Aristotle accepted the four elements but added a version of Plato's Forms as a fifth "essence" that gives a material thing its "soul," makes it more "complete," and brings it closer to the celestial world, where objects move in perfect circles. In Aristotle's logic there were four "causes" of change: formal, material, efficient, and final.

Aristotle also provided a definition of a homogeneous substance, and recognized that ice crystals, liquid water, and water vapor constitute the same substance. He questioned the atomism of Empedocles and Democritos on the ground that it could not account for the difference between a mixture of discrete substances and the combination (*mixis*) of elements to form compounds. Aristotle defined "element" as a "body into which other bodies may be analyzed, present in them potentially or in actuality (which of these is still disputable), and not itself divisible into bodies different in form" (*De Caelo*, 302ª). In his theory, *mixis* is a process in which the "quality" of the participating elements is altered, and that, in the compound, "the elements (are) combined in a determinate mode or ratio" (*De Anima*, 410ª). [11] Of particular interest in relation to Aristotle's influence on later natural philosophers and alchemists was his theory of material change. In his *Metereologia*, he used the term *pepsis* (later translated as *coction* or *concoction*) to denote the conversion of some kinds of wet matter by its "innate (or vital) heat." [12] The word "pepsis" was taken from the Hippocratic writings, where it was associated with the digestion of food. Aristotle extended its meaning by applying it to natural processes that lead to the "perfection" (or "elevation") of the matter undergoing change. Among his biological examples were the ripening of fruit and the development of the animal embryo. He also applied "pepsis" to the action of heat in cooking, and the term was later used to denote the transformation of inanimate objects into living things in the earth.

Although, in his *Sceptical Chymist*, Robert Boyle excoriated the Aristotelian "Peripatetics" who still adhered to the four (or five)-element doctrine in variously-modified form, or the chemists who accepted the salt-sulfur-mercury doctrine of Paracelsus (of whom more later), his definition of an element resembles that offered by Aristotle:

> I now mean by Elements, as those Chymists that speak plainest do by their Principles, certain Primitive and Simple, or perfectly unmingled bodies; which are not made of any other bodies, or of one another, are the Ingredients of which all those call'd perfectly mixt Bodies are immediately compounded, and into which they are ultimately resolved. [13]

Later, Lavoisier wrote: "if we apply the term *elements*, or *principles of bodies*, to express our idea of the last point which analysis is capable of reaching, we must admit, as elements, all the substances into which we are capable, by any means, to reduce bodies by decomposition." [14] Aristotle's definition, with its reference to "analysis," attests to his reliance on the experimental results of craftsmen who decomposed materials by combustion.

Aristotle's views on the general properties of material substances were developed by his pupils, especially Theophrastos, adopted by Muslim scholars (notably Geber, Rhazes, Avicenna), and modified by St. Augustine and by thirteenth-century European theologians and natural philosophers to conform with Christian doctrine. There had been, however, other ancient Greek schools not subservient to the ideas of Aristotle (or Plato), notably that of Epikouros, who based his atomic philosophy on that of Democritos, and these ideas were made known to the Romans during the first century B.C. by Lucretius, in his famous poem *De Rerum Natura*. [15] A fourth Athenian school, founded in about 300 B.C. by Zeno, became known as the Stoics. They tended toward skepticism, rejected atomism, and used the term *pneuma* (or *psyche*) to denote the spiritual agent that gives life to natural things. [16] In first-century Rome, the Stoic philosophy, with an admixture of Epicureanism, was propounded by Seneca, the tutor of Nero. At that time, the major center of learning (also industry and commerce) in the Mediterranean area was still Alexandria. Before the Roman conquest of Egypt (30 B.C.), in addition to the continued influence of these natural philosophies, new schools (Neo-Pythagorism, Neo-Platonism) were created, with the infusion of the religious beliefs of Egyptians, as well as those of the many Greeks, Jews, and Persians who formed a large part of the population. It was during this period that Pliny [17] and Bolos of Mendes collected the available knowledge about such technical arts as dyeing and metal-working, as well as the occult practice of the transmutation of base metals into gold. Although the writings of Bolos were often cited in later papyri, they have not been found, and it can only be surmised that they represent the first descriptions of alchemical practice.[18]

An offshoot of the ancient practical activity of physicians (who often were priests) in preparing medicines and of craftsmen in various technical arts, as summarized by Pliny, alchemy acquired its special character from the addition, to the ideas of Democritos, Plato, and Aristotle, of the mystical beliefs of various religious sects. In about A.D. 300 Zosimus provided a compendium of both the occult and practical aspects of alchemy, with descriptions of the apparatus and ovens used by alchemists.[19] Also, at about that time there appeared the so-called "Hermetic" books, in which alchemy and religious belief were closely intertwined; the thrice-great Greek god Hermes was credited with having invented the arts and magic.

During the years after the rise of Islam, alchemical books written in Greek or Coptic were translated into Arabic. One of the first Muslim scholars to be identified as an alchemist was Jabir ibn Hayyan (latinized as Geber).[20] He considered metals to be composed of "sulfur" and "mercury," and thought that gold, the most "perfect" metal, was a combination of completely "pure" forms of these substances.

According to Geber, transmutation was possible only when a proper balance of the "qualities" of the constituents had been achieved, and he used numerology to define these qualities for various metals. Later, the great Muslim physician Abu Ali ben Sina (Avicenna) adopted many of Geber's ideas, but expressed disbelief in the possibility of transmutation.[21]

After the translation of the Arabic writings into Latin, the alchemy inherited from these sources and developed in Europe during the six centuries before 1800 represented three lines of endeavor. (1) The search for a "philosopher's stone," which would effect the transmutation of base metals into gold, and an "elixir" or "quintessence," which would also serve as a curative agent in human disease. This aim seems to have been the one most encouraged by emperors, kings, and noblemen. The descriptions of the operations were shrouded in secrecy and full of arcane terms and symbols, not only because of competition but also for fear of being accused of magic; the Church guarded its miracles jealously. There was much opportunity for fraud by charlatans who simply colored zinc with sulfur, or "seeded" the initial brew with a pinch of gold. (2) In addition to (or as a part of) the search for the "philosopher's stone," many medieval alchemists conducted experimental studies on the properties of natural substances, provided knowledge about metals, salts, acids and alkalis and solvents (notably alcohol), discovered ammonia and phosphorus, and improved older methods for "separating the pure from the gross" by means of distillation and other laboratory operations. That knowledge represented the background for the emergence of a more reliable technical "chemistry" during the sixteenth century, in the writings of Agricola, Biringuccio, Ercker, and Libavius.[22] (3) An effort was made to develop a natural philosophy in accord with Christian doctrine, as formulated by thirteenth-century clerics such as Albertus Magnus and Thomas Aquinas. In the writings of Robert

Grosseteste, his pupil Roger Bacon and, in the next century, of William of Ockham, Jean Buridan, and Nicole Oresme there are signs of skepticism, emphasis on verification of all experience, and independent investigation of material and efficient causes.[23] The sixteenth-century Reformation altered the dogma somewhat, but the theological influence on chemical thought is evident in the writings of Robert Boyle and Isaac Newton.[24] The mystical aspects of alchemy later attracted Johann Wolfgang von Goethe, Samuel Taylor Coleridge, members of the so-called *Naturphilosophie* movement led by F.W.J. Schelling and, in the twentieth century, the psychoanalyst Carl Jung.[25]

The first of the above objectives of European alchemists—transmutation—died slowly during the eighteenth century. The second led to lines of investigation by men such as Johann Rudolph Glauber and Johann Kunckel who continued the laboratory practices of the alchemists with improved apparatus and discovered (or prepared) many previously unknown chemical substances.[26] During the latter half of the eighteenth century that tradition, without much of the alchemical mysticism, was continued by several French pharmacists and reached a high point in the work of Carl Wilhelm Scheele. The third was transformed during the sixteenth century by an upsurge of Neo-Platonic and Hermetic occultism linked to iatrochemistry and promoted by the Swiss physician called Theophrastus Bombastus von Hohenheim or Paracelsus and his many adherents, notably Johannes Baptista van Helmont.[27] They opposed the humoral doctrine of Galen and advocated the use of particular chemical medicines they considered to be "specific" for particular diseases.[28] The Paracelsians had little use for the ideas of Francis Bacon,[29] whose writings include anticipations of the seventeenth-century mathematics-based "mechanical" philosophy of René Descartes, Robert Boyle, Isaac Newton, and

Robert Hooke.[30] The impact on chemical thought of the so-called Scientific Revolution initiated by Kepler and Galileo, with Newton as its chief luminary, is less clear than that on astronomy and cosmology. Mathematics was not a prominent feature of medieval alchemy. Nor was it prominent in Boyle's and Newton's time in the discourse about the nature of matter.[31] In particular, there was dispute (especially with Thomas Hobbes) about the status of "occult qualities" such as magnetism or gravity, which could not be apprehended by the senses. As godly men, the English advocates of the mechanical philosophy found it necessary (or expedient) to justify experimentation with new instruments as a way to investigate these "qualities" conferred upon natural substances by God. In his *Of the Excellency and Grounds of the Corpuscular or Mechanical Philosophy*, Boyle wrote:

> But when I speak of the corpuscular or mechanical philosophy, I am far from meaning with the Epicureans, that atoms, meeting by chance in an infinite vacuum, are able of themselves to produce the world, and all the phaenomena; nor with some modern philosophers, that supposing God to have put into the whole mass of matter such an invariable quantity of motion, he needed to do no more to make the world; the material parts being able by their own unguided motions, to cast themselves into such a system (as we call it by that name) but to plead only for such a philosophy, as reaches out to things purely corporeal, and distinguishing between the first original of things, and the subsequent course of nature, teaches, concerning the former, not only that God gave motion to matter, but that in the beginning he so guided the various motions of the parts of it, as to contrive them into a world he designed they should compose (furnished with seminal principles and structures, or models of living creatures) and established those rules of motion, and that order amongst things corporeal, which we are wont to call the laws of nature.[32]

In the explanations of natural phenomena such as heat, light, magnetism, or electricity, the mechanical philosophy of Boyle and Newton, based on the motion of "corpuscles" of various size and shape, was riddled with references to "spring," "active principles," "spirits," "subtle fluids," and "aether." In the "General Scholium" of his *Principia*, Newton wrote about

> . . . a certain very subtle spirit pervading dense bodies and lying hid in them, by whose force and actions the particles of bodies attract each other at the very smallest distances, and which when they are brought into contact causes them to cohere; and by this spirit electrified bodies act at greater distances, both attracting and repelling little objects in their vicinity; and light is emitted, reflected, refracted, and inflected; and bodies grow hot; it stimulates sensation, and causes the limbs of animals to be moved at will, for the vibrations of this spirit travel through the solid fibres of the nerves from the external organs of sensation to the brain, and from the brain to the muscles.

Shortly before the appearance of the *Principia*, John Keill and John Freind attempted to put chemical phenomena on a mathematical basis by applying the principle, and some years earlier John Mayow interpreted his experimental results on combustion and respiration in terms of the presence of a "nitro-aerial spirit" and the view that "a vital, igneous and highly fermentative spirit" was "fixed" in saltpeter.[33] (Fig. 1).

Newton's enormous reputation at the turn of the seventeenth century made his views on the nature of matter a part of the preconceptions of several generations of chemists, and it seemed that, because of its explanatory power and simplicity, the Newtonian approach should be applied to the study of chemical phenomena. Indeed, some recent historians, notably Thomas Kuhn and Marie Boas, have considered Boyle's corpuscular theory to be part of seventeenth-century mechanical philosophy.[34] Newton adopted features of the

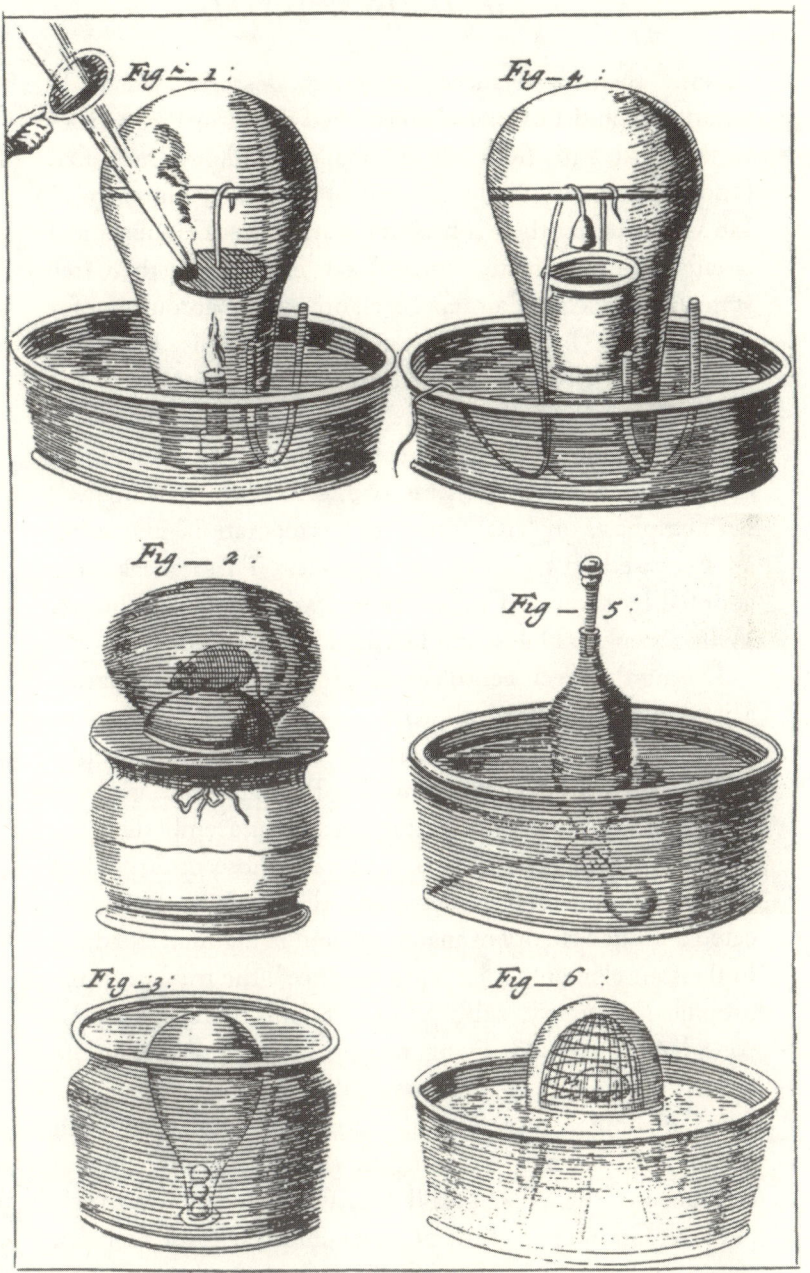

Figure 1. Apparatus for study of gases (Mayow, 1674).

"atomic" theories advanced by Sennert, Descartes, Gassendi, Charleton, and Boyle, and considered the ultimate particles to be "hard, soft, fluid, elastic, malleable, dense, rare, fixt, [and] can be very easily agitated by a vibratory motion." [35] He assumed that their cohesion to form chemical substances is effected by "attractive forces." For example, in their free state, the particles of air repel each other, thus accounting for its "elasticity," but when air is taken up by "dense bodies" it loses its elasticity because the attractive forces overcome the forces of repulsion. Although Newton's chemical reasoning resembles his mathematical reasoning about the movement of planets, it is based on his strong interest in alchemy,[36] with emphasis on "vital forces" or "fermentation" in causing an increased agitation of the particles. These ideas were adopted by several of his contemporaries, notably Thomas Willis, Stephen Hales, and Herman Boerhaave.[37]

During the seventeenth century, some physicians, pharmacists, mineralogists, and alchemists preferred the ideas of Paracelsus to those of Descartes.[38] The arcane philosophy of Paracelsus, based largely on Neo-Platonism and Gnosticism, was centered on medical practice and embodied the notions of earlier alchemists, especially those of Arnald of Villanova, John of Rupescissa and Ramon Lull. He advocated a unified theory of inanimate and living matter, added to the four elements (plus "quintessence") the triad of "philosophical" mercury, sulfur, and salt (plus "Archeus"), and considered that, like living things, minerals arose in the earth from "seeds." [39] Paracelsus also wrote of the "specificity" of human diseases and the need to treat them with "specific" medicines, including those derived from mineral substances, which he is said to have administered to his patients. Coming at a time of social upheaval in the German states, and the inability of traditional medical practice to meet the challenge of epidemics, his aggressive and unconventional behavior, like that of Martin Luther, won Paracelsus considerable popular support, and his denunciation of Galenic medical practice has been compared to Luther's attack on the sale of indulgences. Although later historians have judged his scientific and medical contributions unfavorably, there can be no doubt that Paracelsus attracted many followers, some of whom played significant roles in the development of chemical thought and practice, and their medical applications. Chief among these "Paracelsians" was Johannes Baptista van Helmont; the others included Daniel Sennert, Angelo Sala, Joseph du Chesne, Oswald Croll, François de la Boé (Sylvius), Otto Tachenius, and Johann Joachim Becher.[40] Most of them dismissed the more mystical features of his natural philosophy, and some of them appear to have anticipated Boyle's "corpuscles" by reviving the theory of "Geber" that natural substances are composed of uniform very small particles, while others improved older methods of solution analysis (mainly color tests), which Boyle described in his writings.[41]

As for the problem of "cohesion," which Newton discussed in terms of his theory of universal attraction, the Paracelsian "iatrochemists" referred to the Hipppocratic idea of the action of "vital heat" on moisture to confer upon matter a "glutinous" or "unctuous" quality, as in the kneading of bread. In the writings attributed to "Geber," it was asserted that this quality effects combustion, and some Paracelsians associated it with the philosophical "sulfur." The problem of cohesion also arose in connection with William Gilbert's explanation of the ability of rubbed glass to attract light objects such as feathers. He assumed that the glass emits an electrical "effluvium." This idea was adopted by Kenelm Digby and Thomas Browne, who used the alchemical notion of "unctuosity" to suggest that the effluvium consists of fatty sulfureous particles. Although accepted by Boyle, this explanation could not be fitted into his mechanical philosophy.[42]

The most highly regarded of the Paracelsians, van Helmont, of whom Boyle thought well, was an able chemical worker who used the balance in quantitative experiments but, like Paracelsus, disdained the use of mathematics. He also opposed Galen, and was reported to have been a good physician. He rejected both the four elements and the mercury-sulfur-salt triad as the basic constituents of matter, and asserted that the true elements are air and water. Among his many experiments on this question is the famous one in which he watered a growing willow tree for five years.[43] Van Helmont introduced the word "gas" (or "blas") to denote the invisible exhalation produced during the fermentation of grapes or combustion of wood, which he named "gas sylvester," and he also thought that he had discovered a universal solvent (*Alkahest*).[44] Although he rejected much of Paracelsus's mystic natural philosophy, van Helmont retained the *Archeus* as the animating principle, and considered its function to be the activation of specific "ferments" that effect chemical processes in different organs of the human body. He also wrote of *semina rerum* (invisible and spiritual "seeds") responsible for generation of all natural bodies. Among his disciples were George Starkey (who wrote under the name of Eirenaeus Philalethes) and John Webster. They offered a corpuscular version of van Helmont's natural philosophy.[45]

In his book *Metallographia*, Webster described the generation of metals as a process in which

> . . . the Water being sharp and salt . . . doth meet with the drie, sulphureous, and warm steams that rise from the lower parts of the Earth, do joyn together, and so become unctuous and fat, which settling in close holes . . . as in a close womb, is in length of time thickened into a soft substance,

which they call *Gur*, and after by the warmth of the place, or womb, and its own internal fire, sulphur, or heat, is concocted into a metallick body, pure or impure.[46]

This quotation encapsulated many of the preconceptions inherited by seventeenth-century alchemists, including Boyle and Newton, from sources ranging from Aristotle and the Sophists to the Neo-Platonists. In the formulation of his own mechanical philosophy, however, Descartes offered an entirely different explanation of the evolution of metals, based on the assumption of direct contact of particulate elements, of differences in their size and shape, and of their entry into the interior of existing matter. For example, he considered mercury to be formed from large, heavy, and rounded elements.[47]

This theory attracted wide interest, and led Niels Stensen (usually named Nicolaus Steno), a seventeenth-century Danish anatomist, to examine crystalline minerals with the same care that led to his achievements in the study of the parts of the human body.[48] A pupil of Thomas Bartholin, who demonstrated the independent existence of the lymphatic system, Steno made important discoveries about the glands that form watery fluids (saliva, tears, etc.), and showed that the heart is only a muscle, not the source of the blood or the "vital spirit." He then turned to geological studies, during the course of which he found that crystals grow by the directed apposition of particles in a surrounding medium at surface planes of an existing crystal. This observation antedates, by a century, the beginning of the science of crystallography as developed by Jean Baptiste Louis Romé de l'Isle and René Just Haüy.[49] It should be noted that Pierre Gassendi, a seventeenth-century natural philosopher and Epicurean atomist, introduced the concept of "molecule" as an assembly of "atoms," and the idea that such molecules aggregate to form larger bodies.[50] The term reappeared in the writings of Herman Boerhaave (of whom more later) and, near the end of the eighteenth century, in those of Haüy, who defined the primitive forms of all crystals as "molécules intégrantes."[51]

What some historians have called the Scientific Revolution of the seventeenth century did not include the demise of alchemy. Medieval alchemy was held in high regard by such notables as Libavius, Boyle, and Newton. As "philosophical" investigators, they performed their operations in secret and distanced themselves from the newer variety of Paracelsian alchemists, whom they considered to be vulgar and grasping.[52]

Chapter Two

CHEMICAL COMPOSITION
AND PHLOGISTON

ॐ

DURING THE eighteenth century, the mechanical philosophies of Descartes and Newton had a lasting influence on philosophical thought, but the development of chemical theory and practice proceeded largely through the application of other styles of thought and experiment. To see that development in proper perspective, I will summarize, in admittedly oversimplified fashion, the main lines of technical knowledge gained during the preceding centuries of experimental chemical effort in the Near East and Europe, as seen apart from its Hermetic, Neo-Pythagorean, Neo-Platonic, or Christian orientations.

By the end of the seventeenth century, hundreds of substances, solids classified as "earths" (metals, minerals, salts, alkalis), liquids such as "waters" or "oils," and vaporous materials ("spirits," "essences") had been identified and named by metal-workers, goldsmiths, pharmacists, and other craftsmen (many of them also alchemists).[1] The astrological symbols for the seven "moving stars" were assigned to gold (a circle), silver, copper, iron, lead, zinc, and mercury, and similar symbols were devised for newly found "elements" and for other known materials, such as soda, potash, limestone, saltpeter, borax, or magnesia.[2] Upon strong heating in air ("calcination"), a metal was converted to a "calx," from which the original metal could be recovered by "reduction" (also termed "revivification"). Combustion of the many materials extracted from mineral, plant, and animal sources produced an emission of vapors that often could be condensed and collected by distillation.[3] Aqueous solutions of some of these vapors resembled vinegar, the "acidic" volatile end product of vinous fermentation. During the course of such distillations, it was observed frequently that, after the emission of water-soluble "spirits," there appeared a water-insoluble oily product that was highly inflammable. Some substances (notably sal ammoniac) sublimed on the sides of the distilling flask. What was left in solid form at the bottom of the flask was said to represent "fixed" material or an "earth." Mixtures of an acidic "spirit" and an "earth" produced, upon distillation, another spirit; with sea-salt, the "spirit of salt" was collected in the receiving flask. In many experiments, mixtures of two substances in solution produced precipitates that, after collection and washing, sometimes turned out to be new substances, or having properties identical with, or similar to, those of known substances. The definition of these properties depended on the senses: sight (color, crystalline shape), taste (sour, sweet, salty, pungent), smell (odor), and touch (oily, granular), as well as inflammability, volatility, solubility in various solvents ("menstruums"), and response to tests deemed to be specific for a particular substance or class of substances. For example, acids were known to cause an effervescence when added to some salts, and to turn blue flowers red. Boyle expended much effort in checking the reliability of reported color tests and devised some new ones.[4] The balance was also used, especially by assayers of gold and silver, and it was also known that the two can be separated by means of "aqua regia," a mixture of the "spirit of salt" and the "spirit of nitre."[5] Salts were usually purified by crystallization.

Thus, the decomposition of a chemical material by combustion represented the chief method of "dry"analysis (in the sense of the separation of the parts) and, together with some "wet" methods of solution testing, may be considered to mark the beginnings of what is now "analytical chemistry." [6] Likewise, the formation of new products upon the interaction of two known substances has come to be known as "synthesis." In 1723, Georg Stahl defined "chemistry" as "the art of resolving mixt, compound, and aggregate Bodies into their Principles; and of composing such Bodies from their Principles," and some years later G. F. Rouelle defined it as a "physical art which, by means of certain operations and instruments, teaches us to separate the various substances which enter into the composition of bodies, and to recombine them again, either to reproduce the former bodies, or to form new ones from them." [7]

I have not translated the seventeenth century chemical nomenclature into modern terms because such translation was one of the principal fruits of the experimental efforts of eighteenth-century chemists whose preconceptions and individual styles of reasoning owed more to the writings of the followers of the despised Paracelsus than to those of Descartes and Newton, and because at root it was the accumulated technical experience of chemical craftsmen which defined the problems for experimental investigation. [8] From such experience, chemists learned that it is essential to use materials of the highest attainable homogeneity. Therefore, as Boyle emphasized, improvements in the methods for the purification of a chemical substance are likely to lead to more reproducible experimental results.

For some philosophically-minded historians of chemistry, most of the eighteenth century was the age of the phlogiston theory, formulated by Becher and Stahl, and overthrown by Lavoisier. [9] Much has been written about that theory and about the reluctance of some of Lavoisier's noted chemical contemporaries (especially Priestley, of whom more later) to become "anti-phlogistians." The idea that there exists a "principle" of inflammability, termed "sulfur," and associated with oily ("unctuous") matter, was part of alchemical doctrine. Stahl elevated the principle to the status of a material substance that is lost by a metal such as lead when it is converted to a calx, and regained when the calx is "reduced." It was long known that, upon calcination, metals gain in weight, and there then ensued a lengthy debate about the "levity" or "negative weight" of phlogiston, as well as its role in the emission of heat and light during combustion, and in the electrical "revivification" of metals from calxes.[10] The debate also brought to the fore the question of the definition of an "element," mentioned earlier in this essay. At mid-century, Macquer wrote that in analysis (decomposition)

> In whatever way we attempt to go further, we are always stopped by substances in which we can produce no change, which are incapable of being resolved into others. . . . To these substances we may . . . give the title of Principle or Element. . . . Of this kind the principal ones are Earth, Water, Air, and Fire.[11]

By that time, however, a marked change in the preconceptions of chemists about these four metaphysical principles had become evident, especially in regard to the composite nature of common air, and to the possibility of isolating as "earths" and "waters" unique individual substances of definite composition.

Van Helmont's view that there were "gases" different from common air, and that a particular "gas sylvestre" is emitted during combustion and vinous fermentation was well known during the seventeenth century. John Mayow had devised ingenious glass devices for collecting the "air" emitted during combustion (effected by a "burning lens") or animal respiration. He was followed during the 1720s by Stephen Hales, who invented an apparatus ("pneumatic trough") in which the setup for generating a gas was separated from the one

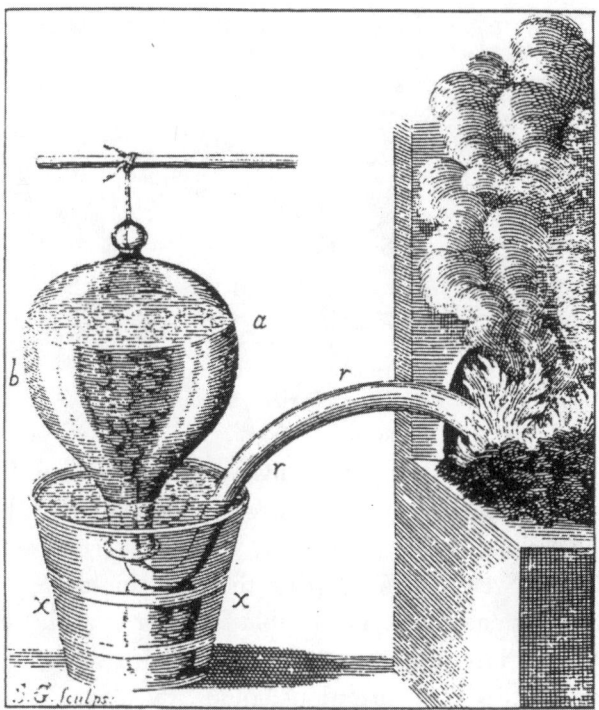

Figure 2a. Pneumatic trough (Hales, 1727).

in which it was collected.[12] (Fig. 2a). In his *Vegetable Stat-icks*, Hales reported quantitative data for the weights and volumes of the "airs" generated upon heating various min-eral, plant, and animal materials, and concluded that "air" is bound in a "fixed" state in many substances. Since he does not appear to have attempted to characterize chemically the "airs" generated from different sources, one must agree with Partington's comment that "if he had paid less attention to Newtonian 'staticks' and more to Mayow, he might have gone much further."[13]

   One of the first of Hales's contemporaries to repeat some of his experiments and to adopt the idea that "air" is retained

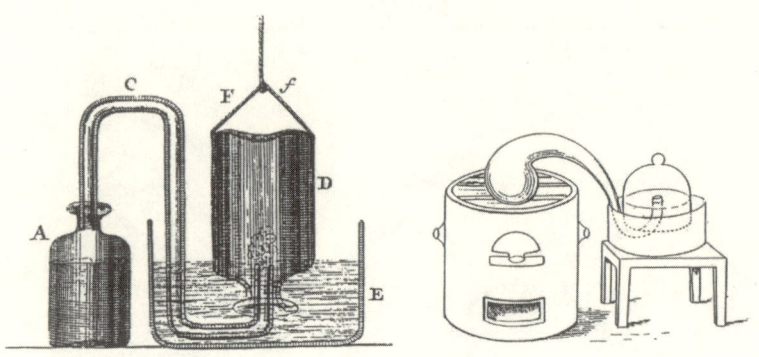

Figure 2b. Apparatus for collecting gases
(left, Cavendish, 1766; right, Lavoisier, 1775).

in an "inelastic" form in many substances was Herman Boerhaave, the famous professor of medicine, botany, and chemistry at Leiden.[14] His *Elements of Chemistry* (1732) and its
translations were for many years the most highly regarded
textbooks of chemistry and notable for their relative lack of
speculation. His personal research dealt with subjects such
as the solubility of air in various liquids. He also offered the
theory of an all-pervasive "fire (*ignis pabulum*)." [15]

The proof of the idea that an "air" is "fixed" in some natural materials was provided by Joseph Black. He did not use
the "pneumatic trough" in his famous work on the composition of magnesia alba, but succeeded in isolating the "fixed
air" released upon the calcination of common alkalis and
alkaline earths such as limestone and magnesia. This discovery led Black to explain the increased "causticity" of a mild
alkali as a consequence of the loss of fixed air.[16] Apart from
the elegance of the experiments, and the modest manner in
which they were reported, Black's quantitative data showed
that, during the calcination, there was no addition of a pon-
derable "principle." This work, which reminded chemists of
Van Helmont's "gas sylvestre" and his idea that there are
different "gases," was soon followed by the achievements

of Priestley, Scheele, Cavendish, Rutherford, and Lavoisier in the isolation of the various constituents of common air. Before moving to those memorable years, I return to the beginning of the eighteenth century, and the contributions of Georg Ernst Stahl and his French adherents to the study of the composition of chemical substances.

I mentioned above that Stahl had defined chemistry as the art of resolving bodies into their "principles" and combining these principles to form compounds. He rejected the Cartesian explanation of chemical phenomena in terms of hooks, sharp points, and wedges, and the Newtonian explanation in terms of intercorpuscular attraction, and preferred the approach described by Agricola and Glauber. For example, Glauber had made his "sal mirabile" by the interaction of "spirit of sulfur" with sea salt, and "sal ammoniac" from "sal volatile urinae" and "spirit of salt." He isolated these products, and other salts made in the same manner, in crystalline form.[17] Glauber noted the difference in affinity of the "acids" for a particular "salt." Stahl also adopted the ideas of Johann Joachim Becher (a contemporary of Boyle), who had proposed that all bodies are composed of "water" and three kinds of earthy "principles" that corresponded roughly to the Paracelsian triad of salt, sulfur, and mercury-vitreous earth, inflammable earth, and mercurial (or fluid) earth.[18] Using Becher's idea, Stahl developed a general scheme for the composition of matter:

> All natural Bodies are either simple or compounded; the simple do not consist of physical parts; but the compounded do. The simple are Principles, or the first material causes of Mixts; and the compounded, according to the difference of their mixture, are either mix'd, compound, or aggregate; mix'd, if composed merely of Principles; compound, if form'd of Mixts into any determinable single thing; and aggregate, when several such things form any other entire parcel of matter, whatever it be.[19]

Stahl used this scheme to interpret his extensive experimental studies on salts, in particular many "syntheses" of the kind described by Glauber, and on acids and alkalis. Stahl also invoked the existence of a "universal acid" composed of water and vitreous earth, and considered alkalis to be composed of a salt and phlogiston. His studies on the reduction (revivification) of a metallic calx (calcination) in the presence of charcoal led him to conclude that the principle of metallicity is the same as the one lost on the combustion of charcoal. He made sulfur by burning spirit of vitriol (which he showed to be identical to oil of vitriol) with charcoal, and also demonstrated in this manner the reduction and artificial formation of phosphorus. Stahl interpreted all these processes, including the action of strong acids on metals, as involving the loss or gain of phlogiston, the principle of inflammability, which he considered to be a fatty earth.

Apart from the formulation of the phlogiston theory, Becher and Stahl emphasized the importance of "fermentations," which they considered to be of three kinds: one in which there is effervescence, with alcoholic fermentation and acetous fermentation as the other two kinds. In his biological theories, Stahl espoused a kind of vitalism that assumed the existence of a soul (*anima*) which worked directly on the chemical processes in living organisms.

Several German adherents of Stahl's theories, notably Johann Andreas Pott and Andreas Sigmund Marggraf (who criticized some of Stahl's claims), made valuable contributions; Friedrich Hoffmann worked on mineral waters, and Caspar Neumann on plant constituents. Also, Johann Juncker and Johann Friedrich Henckel helped to clarify Stahl's writings and made them better known, in particular to the French group of chemists working at the *Jardin du Roi* during the first half of the eighteenth century.[20]

The royal botanical garden in Paris was founded early in the seventeenth century. In addition to growing medicinal plants and preparing drugs, it gave instruction in chemistry.[21] Among the pharmaceutical chemists who worked and taught there before 1700 were Nicaise Le Fèbvre (also spelled Le Fèvre), Christopher Glaser, and (briefly) Nicolas Lemery.[22] All three wrote chemistry textbooks, that of Lemery gaining the most approval. After 1690, Wilhelm Homberg, Etienne François Geoffroy, Guillaume François Rouelle, and Pierre Joseph Macquer conducted research there under the auspices of the *Académie Royale des Sciences*. Homberg did valuable work on the composition of salts, and Geoffroy collected the available data (including those of Homberg and his own) on the relative tendency of one substance (in most cases, an "acid") to interact with another substance to form a salt. Geoffroy's "affinity table" ("table des rapports," Fig. 3) was widely used during the eighteenth century and revised to include new data.[23] The Scottish chemist William Cullen (Black's teacher) connected "affinity" to Newtonian attraction, as did the Swedish chemist Torbern Bergman, who later corrected and greatly enlarged the table. The French chemist Louis Bernard Guyton de Morveau attempted to calculate the force of attraction from a study of crystallization.[24]

At mid-century, Rouelle was the chief protagonist of Stahl's theories in France. They had been introduced earlier by Geoffroy and others, but Rouelle's brilliant lectures brought them into greater prominence despite his relatively modest experimental accomplishments. Equally persuasive was his enterprising student Macquer, who composed a massive dictionary of chemistry. In their expositions, the Stahlian doctrine underwent some revision. Thus, Macquer considered phlogiston to be a fixed form of fire, and he translated Stahl's mixts and compounds into *parties constitu-*

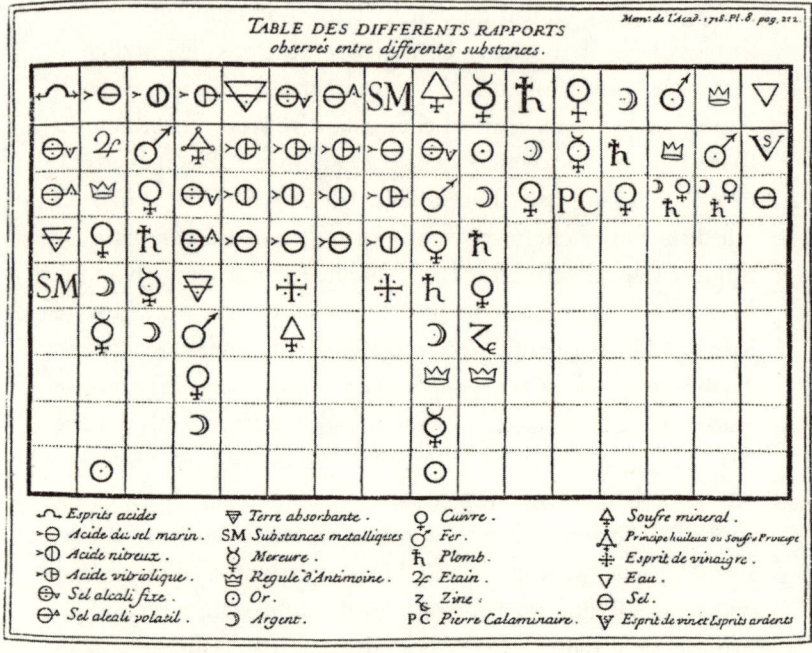

Figure 3. Geoffroy's table of affinities (1718).

antes (for example, the acid and alkali that form a salt) and
parties intégrantes ("the smallest molecule into which a body
may be reduced without being decomposed").[25]

After about 1850, such variants of Stahl's ideas about
chemical composition, and what constitutes the difference
between "simple" substances and "compounds," formed part
of the preconceptions of leading chemists. In retrospect,
the questions regarding chemical composition appear to
have outweighed those related to the existence of "phlogiston."[26] What brought matters to a focus was not a theoretical insight, but the laboratory work of experimenters on
the composition of common air. The essential role of air in
combustion had been demonstrated during the seventeenth
century by Jean Rey, Robert Hooke, and John Mayow, and

Joseph Black's isolation of "fixed air" encouraged the search for other components.

In his celebrated *Traité Elémentaire de Chimie*, published in 1789, Lavoisier wrote of "this air which we—Mr. Priestley, Mr. Scheele and I—discovered at almost the same time, was named dephlogisticated air by the first; empyreal air by the second. At first, I gave it the name *eminently respirable air:* since then I substituted that of *vital air.*"[27] This skillful statement obscures the order of discovery. It appears that Scheele was the first to isolate what came to be called "oxygen," sometime during 1771-1772, by heating various substances (saltpeter, minium, etc.) with pyrolusite ("black magnesia"), which he considered to attract phlogiston strongly. He prepared the new gas in relatively pure form in various ways, and named it "fire air" because it supported the flame of a candle more effectively than common air.[28] With sea-salt, pyrolusite produced a greenish corrosive gas ("dephlogisticated acid of salt"), later named "chlorine." Scheele was also a co-discoverer, with Daniel Rutherford, of "foul air" or "aer malignus," which is lethal and extinguishes flames; it was named "phlogisticated air" by Priestley, and "azote" by French chemists. Among Scheele's numerous other achievements, of special importance was his development of a general method for the isolation, in crystalline form, of acids present in milk (lactic acid) and in plant juices—tartaric acid, citric acid, oxalic acid, malic acid—by first forming the water-insoluble calcium salt, and then decomposing it with a stronger acid. These substances were to appear prominently in the nineteenth-century development of organic chemistry. It must be noted that these achievements were those of an assistant to several apothecaries who gave Scheele time and space for his chemical research, who gained the respect of university professors (especially Torbern Bergman), and who was elected in 1775 a member of the Royal Academy of Sciences of Sweden.[29] It is also noteworthy that although Scheele's acceptance of the phlogiston theory affected his interpretation of the discoveries he made through exceptional technical skill and his imagination, it did not inhibit his research, which laid the groundwork for the more widely recognized accomplishments of others.

The elevation of Lavoisier to scientific sainthood by some nineteenth-century French chemists has recently been questioned by historians (of which more later), as have the parallel denigrations of his chief scientific adversary, Joseph Priestley, [30] who declined to join the anti-phlogistic revolution to the end of his life, a productive life whose last decades were full of turmoil and disappointment. Adopted as a boy by his aunt, he acquired from her the Nonconformist attitude to the Church of England, and became a Unitarian minister. His meager income in that capacity obliged him to become a teacher as well. Although he did not attend a university, he was learned in classical languages, history, the law, and the sciences. In 1767, the year he took up a ministry in Leeds, his first scientific publication, *The History and Present State of Electricity*, appeared—an effort encouraged by Benjamin Franklin, then residing in London. After a seven-year association with Lord Shelburne as his librarian-companion, and who not only provided Priestley with a laboratory but also gave him a life pension, he accepted a ministry in Birmingham, where he was supported by Josiah Wedgwood and others, and had a well-equipped laboratory. As an outspoken religious Dissenter and supporter of the French Revolution, Priestley's opinions were not welcome, and his comfortable life was cut short in 1791, when a mob destroyed his library and laboratory and forced him to flee.[31] In 1794 he emigrated to the United States, where his religious and political views were no more warmly received than in England, but he lived out his last years comfortably, because of the continued generosity of his English friends.

Priestley began his chemical investigations on "airs" in about 1768, using various modifications of Hales's "pneumatic trough" and collecting some of them over mercury. By 1774 he had identified a sizable number of different airs, which he distinguished from each other by such proper ties as solubility in water, color, or inflammability. He also introduced a "nitrous air test," in which a test air was mixed with "nitrous air" over water or mercury, and the reduction in its volume was measured. In 1771, upon heating nitre, he observed the generation of an air that supported the flame of a candle and, as an adherent of the phlogiston theory, considered it to be "dephlogisticated air"— common air from which phlogiston had been removed. By the same token, he named the air unfit for combustion or respiration "phlogisticated air." After innumerable experiments, early in 1775 he recognized (with much "surprise") that an air similar in properties to what he had called "phlogisticated nitrous air" was generated upon heating with a "burning lens" a highly purified sample of the material known as *mercurius calcinatus per se* (also termed "red precipitate") obtained by heating mercury in air. This "air" not only supported the flame of a candle, but repeated experiments with mice demonstrated that it supported their respiration, and experiments with plants showed that they emit this "air," thus "purifying" the atmosphere.[32] The last-mentioned observation stimulated the successive researches of Jan Ingen-Housz, Jean Senebier, and Nicolas Théodore de Saussure on what came to be called "photosynthesis."[33]

In addition to the "phlogisticated nitrous air," Priestley also identified distinct airs, which he named "nitrous air" and "nitrous vapour." The first of the three reappeared in the nineteenth century as an anaesthetic (nitrous oxide, N_2O), the second (nitric oxide, NO) was found during the 1980s to be an important metabolic regulator, and the third (nitrogen dioxide, NO_2) is a useful industrial oxidizing agent.

I will return to Priestley in connection with his dispute with Lavoisier, who challenged his interpretation of the composition of the "phlogisticated nitrous air," and the controversy during the 1780s about the discovery of the composition of water, which pitted Lavoisier against Henry Cavendish and the inventor James Watt. After coming to the United States, Priestley continued his battle against Lavoisier's antiphlogistic theory, and claimed to have obtained "inflammable air" by burning charcoal. William Cruickshank showed, however, that the product is an oxide of carbon (which Lavoisier had named "oxide de carbone") different from "fixed air." It has been suggested that Priestley's death in 1804 may have been hastened by his repeated exposure to the carbon monoxide generated in his last experiments.[34]

Although Priestley gained considerable prestige during the 1770s for his several fruitful discoveries, he also reported many wrong observations and frequently offered doubtful interpretations of his experimental results. In his adherence to the phlogiston theory, over the years his opinion of the nature of phlogiston shifted from its being the matter of heat, to the matter discharged by the lungs during animal respiration, to its identity with inflammable air. He is reported to have said "when I made a discovery, I did not wait to perfect it by more elaborate research, but at once threw it out to the world, that I might establish my claim before I was anticipated. I subjected whatever came to hand to the action of fire or various chemical reagents, and the result was often fortunate in presenting a new discovery."[35]

In contrast to the seemingly unplanned style of Priestley's chemical research, that of Henry Cavendish on what he called "factitious airs" was pursued in a more systematic and quantitative manner. He published fewer papers, the most important ones on this subject appearing in 1766 and during 1783-1786. The record of much of his wide-ranging experimental activity, which also included extensive work on electricity and heat, remained in manuscript form in his private papers. Cavendish drew his theoretical approach from Newton's *Principia* and the queries in the *Opticks*, but appears to have preferred the "point atoms" of Bošković to Newton's "corpuscles."[36] His experimental skill in chemical operations and in the use of instruments probably exceeded that of Priestley and matched that of Lavoisier. Cavendish was a wealthy member of the English nobility, but untypical of his social class in his lack of pretense and in his independence of mind, and was reported to have been reclusive and frugal.[37]

In his 1766 report, Cavendish described the properties of "fixed air" more fully than Black had done, and established the separate existence of the "inflammable air" generated by the interaction between a strong acid and a metal such as zinc, iron, or tin. He devised better apparatus for determining the weight of gases, showed that they differed in density, and recognized that common air is a mixture of "phlogisticated air" and "dephlogisticated air" in constant proportions. He returned to gases after Priestley and John Warltire took up the study of the observation by Alessandro Volta in 1776 that an electric spark applied to a mixture of inflammable air and common air caused an explosion, and Warltire had noticed that "dew" collected on the inside of the glass container. In 1781 Cavendish carried out a series of quantitative experiments using an improved "sparking" eudiometer, and showed that two volumes of inflammable air combine with one volume of dephlogisticated air and are converted to their own weight of water.[38] His explanation of this important empirical result was based on the prevailing preconception that water is an element. He thought that it can exist either as "dephlogisticated air" or as "phlogisticated water" and when the two forms combine, "neutral" water is formed. The report of this work did not appear until January 1784, after Lavoisier had published his 1783 memoir in which he described his experiments on the combination of inflammable air and "vital air" ("principe oxygine"), and the decomposition of water to these two principles.[39] Lavoisier presented his results as evidence for the view that water is not a "simple substance." In his 1784 paper, Cavendish discussed the merits of the two explanations:

> There are several memoirs of Mr. Lavoisier . . . in which he interely (sic) discards phlogiston, and explains those phaenomena which have usually been attributed to the loss or attraction of that substance, by the absorption or expulsion of dephlogisticated air. . . . According to this hypothesis, we must suppose, that water consists of inflammable air united to dephlogisticated air, and indeed, as adding dephlogisticated air to a body comes to the same thing as depriving it of its phlogiston and adding water to it, and as there are, perhaps, no bodies entirely destitute of water, and as I know no way by which phlogiston can be transferred from one body to another, without leaving it uncertain whether water is not at the same time transferred, it will be very difficult to determine by experiment which of these opinions is the truest; but as the commonly received principle of phlogiston explains all phaenomena, at least as well as Mr. Lavoisier's, I have adhered to that.[40]

Cavendish also noted that "another thing which Mr. Lavoisier endeavours to prove is that dephlogisticated air is the acidifying principle. From what has been explained it appears, that this is no more than saying, that acids lose their acidity by uniting to phlogiston, which with regard to nitrous, vitriolic, phosphoric, and arsenical acids is certainly true. . . . But as to the marine acid and acid of tartar, it does not appear that they are capable of losing their acidity by any union with phlogiston." [41] This recognition of a deficiency in Lavoisier's oxygen theory of acidity antedated the demonstration by Humphry Davy in 1810 that "marine acid" does not contain oxygen.

During the nineteenth century there arose the "water controversy" over the priority in the determination of the composition of water. Some historians thought that it was James Watt, who suggested that water is composed of "dephlogisticated air" and "inflammable air," which he considered to be identical with phlogiston. Others assigned the credit to Priestley, with whom Watt had discussed the possibility of converting water to air, and who came to that view after a series of indecisive experiments. Extensive discussion of the way in which information about the work of Cavendish reached Paris led to a verdict in his favor.[42] The historical accounts refer to the "synthesis" of water, but I found no mention of that word (or synthèse) in the 1783 and 1784 papers of Lavoisier and Cavendish. Perhaps the word was not used by Lavoisier because of its Stahlian origin, and Cavendish still considered water to be an "element."

Cavendish also conducted extensive quantitative experiments on the nature of heat. He adopted Newton's view (also that of Francis Bacon) that heat "consists in the internal motion of the particles of bodies," rather than the widely accepted idea that it is an imponderable fluid substance. Nor did he accept Lavoisier's concept of "caloric." Using the best thermometers then available (his father had received a Copley Medal of the Royal Society for the invention of an improved thermometer), Cavendish defined quantitatively the concept of "capacities for heat of bodies" (specific heat). This work was not published in his lifetime, and the credit for laying the foundations of thermochemistry has rightly been given to Lavoisier and his colleague Laplace, who designed an apparatus later known as an "ice calorimeter."[43]

Chapter Three

ANTOINE LAVOISIER

❦

IN 1753, THE French Stahlian Gabriel François Venel wrote in the third volume of the *Encyclopédie*:

> It is evident that the revolution which will place chemistry in the rank it deserves, which will put it at least beside mathematical physics; that this revolution, I say, can only be effected by a skillful, enthusiastic, and bold chemist, who finding himself in a favorable position, and skillfully profiting from some happy circumstances, would attract the notice of scientists, at first by noisy ostentation, by a decisive and affirmative style, and then by reason, if his first weapons have cut into prejudice.[1]

Venel may have had in mind a new Paracelsus, but the frequency with which this passage has been quoted by later historians suggests that they saw in the coming of Lavoisier the fulfillment of this "annunciation."

Lavoisier's father was a wealthy lawyer with connections at the royal court. As a boy, Antoine was raised by a doting aunt and received a broad education. His only college degrees were in the law (1763, 1764). In 1769, he became a junior member of *La Ferme Générale*, a private financial consortium charged by the government to handle leases and collect taxes. In 1780 he was advanced to the rank of *Fermier Général*. That relationship, which brought him considerable wealth, was strengthened by his marriage in 1771 to the 14-year daughter of another *Fermier Général*. An enterprising young woman,

Mme. Lavoisier learned English and draftsmanship to translate articles and prepare drawings for her husband's publications.[2] Upon the outbreak of the 1789 revolution, Lavoisier expressed liberal political views, but during the Reign of Terror, despite his scientific eminence, his association with the hated Ferme led to his indictment, along with 27 others (including his father-in-law) as an enemy of the Republic, and to his execution on May 8, 1794. There can be little doubt that if his life had been spared, given good health Lavoisier would have resumed chemical research and again risen to high estate during Napoleon's reign.[3]

In 1763, Lavoisier joined the geologist Jean Etienne Guettard in a mineralogical survey of France, and during the succeeding five years was stimulated by Rouelle and others to study French translations of Stahl's writings. He also undertook an investigation of the action of water on gypsum (plaster of Paris), and of the question whether water could be converted to earth by prolonged distillation.[4] Lavoisier's evident promise, together with the help of influential family friends, won him election in 1768 as an adjunct member of the *Académie Royale des Sciences*, and as a full member a year later, when he was 25 years old.

In addition to his association with the Ferme, Lavoisier was appointed in 1775 as an administrator of the government *Régie des Poudres et Salpêtres*, responsible for the industrial production of gunpowder. In the following year he moved his laboratory to the Arsenal, near the Bastille. There, along with a splendidly equipped laboratory, Lavoisier had spacious living quarters and a large library. It is reported that he devoted six hours of each day to scientific work, and the rest of the day on the affairs of the *Ferme Générale*, *Régie des Poudres*, and the *Académie des Sciences*. He also received prominent guests at the Arsenal; they included Franklin, Du Pont de Nemours, and Rochefoucauld.[5]

Thanks largely to Lavoisier himself, historians have been able to study the development of his ideas not only from his books and published research papers, but also from the many private notes and laboratory records that he retained, and that were deposited in the archive of the *Académie des Sciences*, along with his correspondence and his regular reports on his work.[6] I cannot take full advantage of this admirable scholarly effort, and will discuss only some aspects of the development of Lavoisier's method and style of research and scientific writing, as exhibited in the principal contributions he is usually said to have made to the Chemical Revolution. They are: (1) the introduction of quantitative methods and new instruments into chemical practice; (2) the elucidation of the process of combustion, its relation to animal respiration, and the overthrow of the phlogiston doctrine; (3) the study of composition of sulfuric, phosphoric, and carbonic acid, and the formulation of the theory that oxygen is the acidifying principle; (4) the introduction of the concept of "caloric" as an imponderable material principle of heat; (5) the clarification of the concept of an element as a "simple substance," and the experimental demonstration that water is a compound; (6) the development, with others, of an improved chemical nomenclature. Although he recognized the importance of the problem of chemical affinity, he did not undertake much experimental work on this subject. I may evince skepticism about some interpretations of Lavoisier's claims because, as was noted in connection with the discovery of oxygen, he was not above making misleading statements on matters of priority.

As regards Lavoisier's use of the chemical balance,[7] it has been stated that he assumed the validity of what came to be called the law of the conservation of mass, and that he was responsible for the introduction of the "balance-sheet" method.[8] He was certainly not the first in that respect, for the assumption was made before him by Black and Cavendish, and before them by Boyle.[9] If he had been the first, one might rank Lavoisier with Galileo, whom Hanson and Koyré sought to change from an experimenter to an inspired theorist.[10] In the chapter on vinous fermentation in his *Traité*, Lavoisier wrote:

> This process [*opération*] is one of the most striking and extraordinary of all those presented to us by chemistry, and we have to examine whence comes the carbonic acid gas which is released, and how a sweet body, a vegetable oxide, can transform itself into two so different substances, one of which is combustible, the other eminently incombustible. One sees that to solve these two questions, it is first necessary to know well the analysis and nature of the fermentable body; since nothing creates itself, neither in artificial operations, nor those of nature, and one can take for granted [*poser*] that in all operations, there is the same quantity of matter before and after the operation; that the quantity and quality of the principles is the same, and that there only changes, modifications.
>
> It is this principle upon which all the art of experimentation in chemistry is founded; one is obliged to assume in all of them a true equality or equation between the principles of the bodies one examines, and those that one obtains by analysis. Thus since *moût de raisin* gives carbonic acid gas and alcohol, I can say that *moût de raisin = acide carbonique + alkool*.[11]

Lavoisier then gave numerical values for the oxygen, hydrogen, and carbon content (by weight) of the crystalline sugar that had been fermented and for the alcohol produced, the oxygen and carbon of the carbonic acid, the oxygen and hydrogen of the water, along with the oxygen, hydrogen, and carbon of the residual unreacted sugar and dry yeast, to produce a perfect balance sheet. He concluded with a mathematical touch that

> . . . I could consider the matter subjected to fermentation
> and the result obtained after fermentation as an algebraic
> equation; and in taking successively each of the elements of
> the equation as unknowns, I can derive a value, and thus
> confirm [*rectifier*] the experiment by calculation, and the cal-
> culation by experiment.[12]

As was noted by Arthur Harden, however, the individual
numbers were incorrect.[13] That, in itself, is not surprising
since Lavoisier could not know what the correct values
might be, and he considered his analytical methods to be
sufficiently accurate for his purpose. What is surprising is
the perfection of his balance sheet. A generous explanation
is the fortuitous compensation of errors, but other less chari-
table explanations cannot be excluded.

Frederic Holmes has described Lavoisier's evaluation of the
quantitative data he obtained during the 1770s as follows:

> It is clear that he sought reliability, but not great precision.
> He routinely estimated the magnitude of errors due to small
> losses he could not measure. He aimed for a complete balance
> sheet of all the materials before and after an operation, but he
> did not expect to arrive at measured quantities exactly. If they
> came close enough to support his interpretation of the opera-
> tion he was studying, that was good enough for him.[14]

In the case of his fermentation experiment, Lavoisier denoted
the samples of the sugar he used as starting material and
the alcohol he isolated as a product. His analytical data suggest, however, these samples were impure (perhaps contami-
nated by water), since the values for carbon were much too
low, and those for oxygen and hydrogen too high. Moreover,
Lavoisier appears to have considered acetic acid to be the
only organic by-product in the process.[15] In this case, a brilliant experiment, which marks the beginning of the modern
study of alcoholic fermentation, may have been flawed by
preconceptions arising from Lavoisier's seeming hope to

bring chemistry, as Venel suggested, on a par with mathematical physics.[16]

Like many other younger chemists of his time, Lavoisier was attracted during the late 1760s to processes in which air was taken up or released, as in calcination and combustion or animal respiration. In November 1772 he deposited with the secretary of the *Académie des Sciences* a sealed note in which he proposed the theory that an elastic air is taken up in the calcination of metals and other substances, and in April 1773 he presented his theory at a public session of the Academy.[17] Sometime between the two dates he came to know of the work of Black and Priestley, and much of his experimental effort during 1773 dealt with the question of whether fixed air is the one absorbed during the calcination of sulfur, phosphorus, and metals such as mercury or lead. At one stage, he thought that the "acidum pingue" described by Johann Friedrich Meyer, a critic of Black's theory of causticity, might be the elastic air he was hoping to find, and he also examined Priestley's "nitrous air" as another possibility.[18] In that year, Lavoisier undertook the preparation of his *Opuscules Physiques et Chimiques*, which was published in 1774. The sealed note, which was opened in May 1773, ended with the statement that "this discovery seeming to me one of the most interesting made since the time of Stahl, I thought it my duty to assure myself of priority by depositing this note. . . ."[19] At the end of 1774, however, the status of the relation of Lavoisier's elastic air to Black's fixed air or Meyer's acidum pingue was unclear. According to Holmes,

> By January 1774, Lavoisier had learned some hard lessons. The beautiful rigorous style of demonstration that he admired in geometry would not work in chemistry. The many setbacks he had encountered as he tried to gather evidence to support a brave new theory had taught him that he, too, must follow "another route."[20]

Lavoisier did not refer in the *Opuscules* to the experimental work of the French military pharmacist Pierre Bayen, who had suggested in 1773 that the increase in weight of a metal undergoing calcination is due to the fixation of an "elastic fluid" and had obtained this air by heating minium without the addition of charcoal.[21] As was noted earlier in this essay, Priestley discovered in 1774 that what he called "dephlogisticated air" supported the respiration of a mouse, and thus was different from his nitrous air. According to his account, in the fall of that year he visited Paris and told Lavoisier about his discovery.[22] Lavoisier promptly confirmed Priestley's finding (he used a sparrow) and named the inspired gas "l'air emminement respirable," and the fixed air that had been emitted during respiration, and had given a positive response in Black's limewater test, "acide crayeux aeriforme."[23]

The identification and isolation of the elastic air that supports respiration as a new component of common air was followed by Lavoisier's intensive research on its role in the formation of acids, and the claim in 1776 that "acids are composed in great part of air. . . . Not only air but the purest part of air enters into the composition of all acids without exception, and it is this substance which constitutes their acidity."[24] Hence his decision to rename his "vital air" again to denote its acid-forming capacity. As was mentioned earlier, this oxygen theory of acidity was questioned by Cavendish, and later disproved by Davy. More lasting, however, has been the concept of "oxidation" as a counterpart of the ancient idea of "reduction."

During the 1780s, Lavoisier collaborated with Pierre Simon de Laplace in performing important thermochemical experiments, and in formulating the concept of an imponderable "matter of fire" associated with oxygen. In 1783, he declared that in combustion the combustible material does not release phlogiston but takes up the matter of heat from oxygen.[25] By 1790, what he named "caloric" had replaced phlogiston in the thought of many chemists, especially the younger ones, but versions of the phlogiston doctrine persisted into the early years of the nineteenth century.[26] Kant welcomed the caloric theory because of its seeming similarity to his own concept of a "Wärmestoff";[27] Sadi Carnot accepted the caloric theory in his formulation of a theory of the operation of the steam engine.

Also, during the 1780s, Lavoisier began the preparation of a chemistry textbook; it appeared in 1789 as the *Traité Elémentaire de Chimie*. Most books of this sort were based on lectures given by teachers; this one was different because of its emphasis on Lavoisier's work and thought.[28] A distinctive feature was a table of "simple substances" (Fig. 4), defined as those that had not yet been resolved into simpler "elements" by chemical analysis. The names assigned to many of them were new and had been included in the *Méthode de Nomenclature Chimique*, published in 1787 under the authorship of L. B. Guyton-Morveau, A. Lavoisier, C. L. Berthollet, and A. F. Fourcroy. That collaborative effort to produce an improved chemical nomenclature, with shorter and more informative indications of composition, was based on the earlier writings and correspondence of Torbern Bergman and Guyton-Morveau.[29] The Table in the *Traité* lists light, caloric, azote, oxygen, and hydrogen, which Lavoisier denoted as "simple substances which belong to the three realms [of nature] and which one may perhaps consider as the elements of bodies"; a second group of non-metallic substances (sulfur, phosphorus, carbon, and the "radicals" muriatic, fluoric, boracic), which are "oxidizable and acidifiable"; a third group of 17 metals, which are also "oxidizable and acidifiable"; and a fourth group of

	Noms nouveaux	Noms anciens correspondans
Substances simples qui appartiennent aux trois règnes & qu'on peut regarder comme les élémens des corps	Lumière	Lumière.
	Calorique	Chaleur. Principe de la chaleur. Fluide igné. Feu. Matière du feu & de la chaleur.
	Oxygène...............	Air déphlogistiqué. Air empiréal. Air vital. Base de l'air vital.
	Azote.................	Gaz phlogistiqué. Mofète. Base de la mofete.
	Hydrogène	Gaz inflammable. Base du gaz inflammable.
Substances simples non métalliques oxidables & acidifiables	Soufre	Soufre.
	Phosphore	Phosphore.
	Carbone	Charbon pur.
	Radical muriatique	Inconnu.
	Radical fluorique	Inconnu.
	Radical boracique	Inconnu.
Substances simple métalliques oxidables & acidifiables	Antimoine	Antimoine.
	Argent.................	Argent.
	Arsenic	Arsenic.
	Bismuth	Bismuth.
	Cobolt	Cobolt.
	Cuivre	Cuivre.
	Etain	Etain.
	Fer....................	Fer.
	Manganèse	Manganèse.
	Mercure................	Mercure.
	Molybdène	Molybdène.
	Nickel	Nickel.
	Or	Or.
	Platine	Platine.
	Plomb.................	Plomb.
	Tungstène..............	Tungstene.
	Zinc..................	Zinc.
Substances simples salifiables terreuses	Chaux	Terre calcaire, chaux.
	Magnésie...............	Magnése, base du sel d'Epsom.
	Baryte.................	Barote, terre pesante.
	Alumine	Argile, terre de l'alun, base de l'alun.
	Silice	Terre siliceuse, terre vitrifiable.

Figure 4. Table of simple substances (Lavoisier, 1789).

"salifiable" substances (*chaux, magnésie, baryte, alumine, silice*).
The list did not include the alkalis potash and soda, which had
not yet been resolved into their elements, and which Lavoisier
believed to be compounds.[30]

The second part of the *Traité* is entitled: "On the combination of acids with salifiable bases, and the formation
of neutral salts." This section contains a series of tables
with new names of binary compounds (for example, oxides)
formed by the oxygenation of metals and non-metallic substances, those composed of simple substances and azote
(ammonia), sulfur (sulfuric acid), carbon (carbonic acid),
and the acid radicals, each in their affinity for a series of
bases, with the formation of neutral salts. Lavoisier provides
another example of his style in stating in the preface that
"this second part does not contain anything that belongs
to me (*qui me soit propre*); it only presents a very concise
summary of results obtained from various writings."[31] The
second volume deals largely with the instruments and experimental methods used in Lavoisier's laboratory—the balance, gas analysis, calorimetry, distillation, combustion.

Lavoisier's basic table of "simple substances," which has
been accorded a high place in the history of chemistry, also
reveals something of his style in the presentation of his theories. One is struck by the inclusion of caloric as the matter
of heat in the form of a subtle fluid associated with oxygen,
which exerts a repulsive force on the "molecules of matter."
The emphasis on the oxygen theory of acidity stands in
contrast to Lavoisier's professed intention to avoid speculation. As to Lavoisier's nomenclature, it is of interest that
while Bergman sought the advice of Linneaus about using
the concepts of classes, genera, and species in the naming
of chemical compounds, and about the desirability of using
Latin or an artificial language for this purpose, Lavoisier
cites the writings of the philosopher Condillac as his prin-

cipal guide in matters of nomenclature.[32] As it turned out, Lavoisier's new names for the gaseous elements had a mixed reception abroad. In Germany, oxygen became *Sauerstoff*, carbon became *Kohlenstoff*, and hydrogen became *Wasserstoff*. In Russia, the corresponding names became *kislorod*, *uglerod*, and *vodorod*. Azote became nitrogen in England, and *Stickstoff* in Germany.

Lavoisier's style is also evident in the preface of the *Traité*, where he explains the omission in the text of a discussion of chemical affinity. He states that his self-imposed "rigorous law" to treat only the well-established parts of chemistry

> . . . did not allow me to include in this book the part of chemistry which perhaps is most likely to become an exact science; it is the one which deals with chemical affinities or elective attractions. Mr. Geoffroy, Mr. Gellert, Mr. Bergman, Mr. Scheele, Mr. de Morveau, Mr. Kirwan and many others have already collected a multitude of particular facts, which only await the place to which they should be assigned; but the principal data are lacking, or at least those which we have are insufficiently precise or certain, to become the fundamental basis for such an important part of chemistry. The science of affinities is in any case to chemistry as transcendental geometry is to elementary geometry, and I did not think it my duty to complicate the existing difficulties of the simple and easy elements which, I hope, be within the reach of a very large number of readers.
>
> Perhaps it is a feeling of self-esteem [*amour propre*], without my taking account of it, that has attached importance to these reflections. Mr. de Morveau is at present publishing an article [on] AFFINITY in the *Encyclopédie Méthodique*, and I have much cause to fear working in competition with him.[33]

In connection with this statement, it should be noted that in 1782 Lavoisier presented a *mémoire* on the affinity of oxygen for different substances with which it can combine, and he also criticized the available affinity tables:

What I have just said against tables of affinity in general naturally applies to the one which I am going to present; but I do not think any the less that it can be of some utility, at least until increasing numbers of experiments, and the application of calculation to chemistry, put us into a position to carry our views further. Perhaps, one day, the precision of the *data* will be brought to the point where the geometer will be able to calculate, in his study, the phenomena of any combination whatever, so to speak in the same manner as he calculates the movement of celestial bodies. The views which M. de Laplace has on this project, and the experiments which we have planned, in accordance with his ideas, for expressing by numbers the force of the affinities of different substances, allow us already not to look on this hope as absolutely as a chimera.[34]

So far as I have been able to learn, nothing came of this "project," but these two statements provide additional examples of Lavoisier's style of thought and writing.

In view of Lavoisier's high standards regarding the problem of chemical affinity, I was surprised to find in the *Traité* a chapter on the decomposition of vegetable and animal materials by the action of fire. There he offers some "particular facts" about the vapors and oils produced by the action of boiling water and of "red heat," and states that he will present the details at a meeting of the *Académie*.[35]

As a young scientist, Lavoisier was avid for glory. He made excessive claims for the originality of his theories and experiments, and tended to omit reference to contributions that antedated his own. This tendency was moderated in later years, but even then his assertions of priority were occasionally unjustified.[36] I can only raise the question of the extent to which the style of a *Fermier Général*, experienced in the investment of capital and the assessment of financial risk in a competitive marketplace might have been reflected in the style of Lavoisier's scientific thought, practice, and conduct. Perhaps further study of the archival material dealing with his activities as a financier and an administrator might throw light on some of the contradictions in his behavior as a scientist.

There can be little doubt that Lavoisier himself, and his close associates (especially Fourcroy), believed in 1789 that a Chemical Revolution, comparable to the political revolution of that year, had been effected by the reinterpretation, in terms of the oxygen-caloric theory, of the phenomena previously explained in the phlogiston theory.[37] Lavoisier's letter in 1790 to Benjamin Franklin, inviting him to join the antiphlogistic party, or to Joseph Black in the same year, welcoming the news of Black's conversion, seem to reflect the revolutionary spirit of the time. Nor can there be any doubt that Lavoisier was chiefly responsible for the change in chemical thought and practice, whether it be considered a "revolution" or "reform," or that the chemical discoveries during the preceding years by adherents of Stahl's phlogiston theory were no less important than Lavoisier's rejection of that theory. If Lavoisier, like other chemists of his time, hoped to raise the status of chemistry to a par with mathematical physics, conducted experiments with physical instruments, and even occasionally called himself a physicist (*physicien*), the principal impact of his thought and work was on the development of chemistry, not on physics.[38]

During the nineteenth century, Lavoisier became a symbol of the greatness of France.[39] The adulation accorded him, like that later given to Louis Pasteur, is part of French social and political history, and is largely a consequence of the emergence of Germany as the leading world center in the development of chemical knowledge and its industrial applications.

Chapter Four

ATOMS, EQUIVALENTS, AND ELEMENTS

꒜

THE WIDESPREAD acceptance, during the late 1780s and the 1790s, of the antiphlogistic theory, and the enunciation, in the next decade, of John Dalton's atomic theory, overshadowed the contribution of Jeremias Benjamin Richter to the problem of the proportions in which two chemical substances combine.[1] A pupil of Immanuel Kant in Königsberg, Richter received his doctoral degree in 1789 for a dissertation entitled *De usu matheseos in chymia*, but, unable to secure an academic post, he worked as a chemist in a mining office and then in a porcelain factory.

In a series of papers (1791-1802), Richter provided one of the most striking examples of the mathematical style of chemical reasoning. It was known that when two neutral salts in aqueous solution exchange their acidic and basic components, the solution remains neutral, and Richter performed numerous experiments to determine the numerical values for the combining masses in reactions of this kind. Thus, when calcium acetate and potassium tartrate are mixed, calcium tartrate separates from the solution, and potassium acetate remains in solution. "This is found by experience to hold for all decompositions by double affinity, in so far as the compounds used in the decompositions are also neutral . . . [and that] the elements must, therefore, have among themselves a fixed ratio of mass."[2]

Bases		Acids		
Alumina	525	Hydrofluoric Acid		427
Magnesia	615	Carbonic	"	577
Ammonia	672	Sebacic	"	706
Lime	793	Muriatic	"	712
Soda	859	Oxalic	"	755
Strontia	1329	Phosphoric	"	979
Potash	1605	Formic	"	988
Baryta	2222	Sulphuric	"	**1000**
		Succinic	"	1209
		Nitric	"	1405
		Acetic	"	1480
		Citric	"	1583
		Tartaric	"	1694

Figure 5. Richter's table of chemical proportions
(as collected by Fischer 1802).

Owing to an error by Berzelius in 1819, this approach to the study of what Richter called "stoichiometry" was wrongly credited to Carl Friedrich Wenzel. Although corrected by Berzelius himself, and others, this error continued to be repeated.[3] Richter's mathematical presentation made his writings difficult for chemists, but Ernst Gottfried Fischer helped by setting up a table of "reciprocal proportions" (Fig. 5), which listed, in separate columns, bases with their "equivalent" mass numbers (for example, ammonia, 672) and acids (for example, nitric acid, 1405). The value of 1000 was arbitrarily set for sulfuric acid. The numerical value of the weight of a base required for the complete neutralization of any of the acids was given by the ratio of two such numbers, and likewise the ratio for the neutralization of an acid by any of the bases was given by the reciprocals of such ratios. Richter claimed that the equivalent weights for the bases increased in arithmetical progression, while those for the acids increased in geometric progression. This over-mathematization of his findings was generally disbelieved and contributed to skepticism about the accuracy of his experimental results, as well as to the suspicion that numbers had been adjusted to fit his theory.

Parallel to Richter's studies, the better-known Claude Louis Berthollet, in 1799, approached the problem of chemical proportions from a critical examination of the views of Bergman and Guyton de Morveau on chemical affinity. Along with other chemical colleagues of Lavoisier, Berthollet survived the Terror. In 1794, he was a member of the group charged by the Convention to plan the establishment of what came to be the famous *Ecole Polytechnique*. He was befriended by Napoleon, who took him along on the 1798 expedition to Egypt, where Berthollet composed his first paper on affinity and mass action. He was active in the promotion of the French dye industry, and of the use of "oxymuriatic acid" (discovered by Scheele) as a bleaching agent.[4] In 1807, now one of the scientific elite of the Empire, Berthollet organized the *Société d'Arcueil*, to which many leading French scientists belonged.[5]

Berthollet, five years younger than Lavoisier, studied medicine in Turin and Paris, and chemistry with Macquer, whom he succeeded as inspector of dyeworks. Before 1785, when he became an adherent of the antiphlogistic theory, Berthollet had made valuable contributions to the study of several substances, especially ammonia, and had questioned the validity of Lavoisier's oxygen theory of acidity.[6] In his *Essai de Statique Chimique* (1803), which dealt largely with chemical affinity, he offered an alternative theory of acidity, and in writing of chemical proportions, he recognized the merits of Richter's work and of Fischer's presentation of his data.[7] Berthollet's criticism of the affinity tables assembled by Bergman and Guyton de Morveau stressed the uncertainties implicit in their assumption that, in all cases, the reaction in question had gone to completion. He presented experimental data showing that this assumption is unjustified when all the reactants and products remain in solution, and only applies when one of the products separates from the solution in the form of a vapor or as an insoluble precipitate. Berthollet offered a Newtonian explanation of the concept of affinity, in which various factors, such as temperature, may affect the attraction and repulsion of the molecules of a substance. In particular, he emphasized the importance of the "effective mass" (or "chemical mass") of a substance in determining the direction in which a reaction may proceed, and that in solution there may be an "equilibrium" between the amounts of reactants and products. One consequence of Berthollet's mechanistic approach was his view that substances could combine in variable proportions between the limits of incipient combination and total "saturation." There then ensued his famous controversy with Joseph Louis Proust during 1802-1808. Proust showed that substances formed under very different conditions have a fixed composition, and that the cases cited by Berthollet involved mixtures.[8]

Proust's data were not the only factor in the rejection of Berthollet's idea of variable proportions. A more compelling reason came during the first decade of the nineteenth century, when John Dalton, a Quaker, a teacher, and a member (later president) of the Manchester Literary and Philosophical Society, undertook a different approach to the problem of chemical proportions.[9] His student and biographer William Charles Henry (son of William Henry) wrote of Dalton that (in contrast to Priestley) "he probably never instituted a single experiment without a clearly preconceived object," and Dalton himself stated that "having been in my progress so often misled, by taking for granted the results of others, I have determined to write as little as possible but what I can attest by own experience." [10]

Dalton's chemical atomic theory, as formulated during the years 1802-1808, was based on the assumptions that chemical elements are composed of identical and undecomposable atoms of different masses, and that in chemical reactions the atoms combine in multiple proportions, with whole number ratios such as 1:1, 1:2, 2:3, etc. His "progress" in the development of this theory, with which Berthollet's idea of variable proportions was clearly incompatible, has been the subject of much discussion, especially since the publication in 1896 of *A New View of the Origin of Dalton's Atomic Theory* by Henry Roscoe and Arthur Harden.[11]

Dalton's initial scientific interest was in meteorology, and his first book was entitled *Meteorological Observations and Essays*. He later wrote:

> Having been long accustomed to make meteorological observations, and to speculate upon the nature and the constitution of the atmosphere, it often struck me with wonder how a *compound* atmosphere, or a mixture of two or more elastic fluids, should constitute apparently a homogeneous mass, or one in all mechanical relations agreeing with a simple atmosphere.[12]

Starting with the well-known Newtonian assumption that "elastic fluids" are composed of corpuscles that repel each other, and the newer knowledge that common air contains several "airs" of different specific gravity, in 1801 Dalton considered that the relationship of the individual components in a mixture of gases could be explained in terms of physical repulsive forces by assuming that, in a mixture of two gases, the particles of one gas do not repel those of the other gas. This mechanical hypothesis was quickly challenged by Thomas Thomson, John Gough, and others, and his own growing doubts about the theory led him to determine the relative weights of the atmospheric gases, their solubility in water, and also to include the oxides of nitrogen in his studies. By 1803, with the results of this investigation, and increased acquaintance with earlier reports on chemical proportions (including those of Richter), he had formulated his chemical atomic theory, embodying the "laws" of the conservation of mass, fixed (or definite) proportions, reciprocal proportions, and multiple proportions. Dalton used circles as symbols for the individual atoms; oxygen was denoted by a ◯, hydrogen by a ⊙, carbon by a ●, water by ⊙◯, carbonic acid by ◯●◯. He also presented a table of atomic weights, based mostly on the analytical data of other chemists.

By 1808, when Dalton presented the fully developed form of his atomic theory in the first volume of his *A New System of Chemical Philosophy*, where he also included studies on heat, the theory had received experimental support from the work of Joseph Louis Gay-Lussac [13] on the combination of gases, of William Hyde Wollaston on superacid and subacid salts, and of Thomas Thomson on oxalates.[14] Reservations were expressed, however, by Berthollet and by Humphry Davy, in the latter case partly on the ground that Dalton had been anticipated by William Higgins.[15] The most important endorsement came from Jöns Jacob Berzelius, who stressed the necessity of more accurate determinations of atomic weights. This activity occupied the attention of several generations of chemists in succeeding decades.

Dalton's atomic theory neatly solved the problem of the definition of the *concept* of an element—it is a body composed of identical atoms—but, as Davy noted, it did not solve the practical problem of determining whether a given substance is an element or a compound. The removal of phlogiston from the list of chemical constituents made charcoal (carbon) a "simple substance." But what of the "acid radicals" in Lavoisier's 1789 list? Moreover, during the next 20 years, some 13 new metals were identified and isolated.[16] This problem was attacked most boldly by Humphry Davy.

According to his biographers, Davy's appetite for chemistry

was whetted by the study of Lavoisier's *Traité* and nourished by his association with Thomas Beddoes at the Pneumatic Institution in Bristol.[17] There, Davy discovered the pleasant physiological effects of nitrous oxide, and attempted to replace Lavoisier's concept of caloric with the idea that oxygen is combined with light to form what Davy called "phosoxygen." The rapid demise of that idea apparently convinced Davy not to offer speculations in his later scientific papers. At Bristol, he also wrote poetry, a literary activity then favored by some English scientists, including Beddoes and Erasmus Darwin, and gained the friendship of Robert Southey and Samuel Taylor Coleridge. After Davy moved in 1801 to the Royal Institution in London, Davy distanced himself from Coleridge, whose concept of chemistry became infused with the idealistic German philosophy.[18]

The Royal Institution, founded in 1799 by Count Rumford (Benjamin Thompson), provided a favorable setting for the development of Davy's talents. He was made professor in 1802, and delivered the public scientific lectures, with experimental demonstrations, very effectively. The presentation of such lectures has been a part of the tradition of the Royal Institution throughout its distinguished and still continued life.[19]

Immediately after the reports in 1800 on the Voltaic pile and that of William Nicholson on the galvanic decomposition of water, Davy (and others) undertook systematic studies of the phenomenon of the "electrolysis" of chemical compounds.[20] The most famous fruit of his experiments was the discovery in 1807 of the alkali metals potassium (from potash) and sodium (from soda), two remarkable substances that burst into flame upon coming in contact with water. This finding had immediate relevance to Lavoisier's oxygen theory of acidity and his table of "simple substances." As Davy put it: "Oxygene then may be considered as existing in, and as forming, an element in all the true alkalis; and the principle of acidity of the French nomenclature, might now likewise be called the principle of alkalescence." [21]

As a skeptical student of Lavoisier's *Traité*, Davy used the term "element" sparingly, and preferred "undecompounded body." This attitude was an experimenter's response to the discovery of numerous new metals and, in particular, to the speculation of Gay-Lussac and Thenard that potassium and sodium might be combinations of alkalis with hydrogen. After Davy's demonstration in 1810 that "oxymuriatic acid" did not contain oxygen and was, in fact, another undecomposable and inflammable body he named "chlorine," [22] he expressed the view of many chemists of his time:

> It is contrary to the usual order of things, that events so har-
> monious as those of the system of the earth, should depend on
> such diversified agents, as are supposed to exist in our artificial
> arrangements; and there is reason to anticipate a great reduc-
> tion in the number of undecompounded bodies. [23]

This opinion led Davy to question the connection of "Mr. Dalton's ingenious idea" and the practical principle of definite proportions and, in an attempt to reduce the number of elements, to flirt briefly with phlogiston-like hypotheses. Also, like other chemists of his time, Davy expressed doubts about what came to be called "Prout's Hypothesis." [24]

In 1815, there appeared an anonymous publication in which the author offered a set of tables for the atomic weights of "elementary substances," either as such or in chemical compounds, with that of the lightest one (hydrogen) being set as 1. [25] The author was soon identified as William Prout, a respected London physician, well-known for his skill and precision as an analytical chemist. Among his numerous contributions in what was then called "animal chemistry" were valuable studies on uric acid and the surprising demonstration that the acid of gastric juice is hydrochloric acid.[26]

Like Richter before him, Prout sought to mathematize, and thus to simplify, chemistry. In his 1815 paper, he wrote:

> . . . we may notice 1. That all the elementary numbers, hydrogen being considered as 1, are divisible by 4, except carbon, azote, and barytium, and these are divisible by 2, appearing to indicate that they are modified by a higher number than that of unity or hydrogen. Is the other number 16, or oxygen? And are all substances compounded of these two elements?[27]

Together with the other points made by Prout, his hypothesis (some people later called it "Prout's Law") generated a controversy that lasted throughout the nineteenth century. The precise replicate analyses of many elements (including hydrogen) gave values, within the limits of error, that could not be rounded off to whole numbers, the most convincing case being that of about 35.5 for chlorine. Among Prout's prominent critics over the years were Berzelius, Jean Servais Stas, Mendeleev, and Theodore William Richards, and among his defenders were Thomas Thomson, Jean Baptiste Dumas, Jean Charles Galissard de Marignac, and Lothar Meyer. Stas called the Prout Hypothesis an "illusion," while Marignac praised it as a statement of the "unity of matter."[28] Whatever the judgment of its merits, one cannot but agree that in stimulating the invention of better methods for determining atomic weights, "few hypotheses have been so persistently fruitful."[29] I will consider shortly some of the later developments in connection with the emergence of organic chemistry after 1820.

To return to Davy, he resigned his professorship at the Royal Institution in 1813, but continued his research there. A few years later, he won further renown through studies on flame, and the resulting invention of a miner's safety lamp, in which a lighted candle was protected by a wire gauze. In the latter stages of that investigation, Davy discovered that a preheated platinum wire effected the "slow combustion" of various substances (methane, alcohol, etc.). He did not examine the matter intensively, but the problem was taken up during the 1820s by Johann Wolfgang Döbereiner, who greatly extended Davy's work;[30] this phenomenon was later named "catalysis" by Berzelius. By the 1830s, the concept of catalysis had become linked to the study of "ferments" (it later also assumed great importance in industrial chemistry), and it seems appropriate to postpone the further discussion of that topic.

In the latter years of his life, Davy began to travel frequently on the continent, and his last book, *Consolations in Travel*, was published (by his brother John) in 1830, a year after his death in Geneva. It was during those years that Davy added to his many accomplishments the employment of a blacksmith's son—Michael Faraday—as a laboratory assistant at the Royal Institution.

It will not have escaped the notice of the informed reader that I omitted from the above account mention of the "molecular" hypothesis offered in 1811 by Amedeo Avogadro, and supported by André Marie Ampère in 1814. Based on the demonstration in 1809 by Gay-Lussac that gases always combine in simple proportions by volume and, when the product is a gas, its volume is also simply related to that of its components, Avogadro's hypothesis states that the number of integral molecules in any gases is always the same for equal volumes.[31] No doubt Berzelius knew of the Avogadro-Ampère hypothesis, but chose to ignore it in writing his annual reviews, and in his up-to-date compendious history of chemistry (1843-1847), Hermann Kopp does not even mention Avogadro by name.[32] As the story is often told, it was not until 1858 that Stanislao Cannizzaro reminded chemists of the importance of Avogadro's "forgotten" contribution. In fact, however, it was taken up eagerly for a time by Dumas, and later formed a part of the theoretical basis of the "chemical revolution" initiated by Auguste Laurent and Charles Gerhardt during the 1840s in their approach to fundamental problems of the newly emergent organic chemistry. Apart from leading to the development of the concept of "molecular structure" by Couper, Butlerov, and Kekulé (the subject of a later section), that approach proved to be fruitful in providing new interpretations of the empirical knowledge about metals, acids, alkalis, and salts gained in the past.

Chapter Five

RADICALS AND TYPES

❧

IN EIGHTEENTH-CENTURY textbooks such as those of Boerhaave and Neumann, the many chemical substances isolated
from plant and animal sources were described in sections
separate from those devoted to metals, salts, acids, and alkalis. The products from plants included sugars, camphor, and
dyes (indigo), as well as rubber, tannin, and gums. Those
from animals included egg albumin, bone gelatin, and blood
fibrin. When they were subjected to elementary analysis later
in the eighteenth century, it was found that these products
all contained carbon, oxygen, and hydrogen, and that animal
substances had a higher proportion of nitrogen. This also
appeared to be the case for the many such substances isolated (often in crystalline form) during 1780-1820 by Scheele,
Fourcroy, Vauquelin, Pelletier, Caventou, Braconnot, Prout,
Wollaston, and others. The high point in this development
was reached in the work of Chevreul on the constitution of
fats, with the isolation of glycerol and the various fatty acids
of which the individual fats were composed.[1]

In his multi-volume compendium of chemical knowledge,
Système des Connaissances Chimiques (1801-1802), Fourcroy
included long sections dealing with plant organic compounds
and animal organic compounds, with the term "organic"
to indicate their biological origin, and probably taken from
Buffon. Fourcroy also referred to these substances as "proxi-
mate principles"; others later used the term "immediate prin-

ciples." [2] The study of the composition of these compounds became the province of the specialty named by Jöns Jacob Berzelius "organic chemistry," [3] with animal chemistry allied to physiology and medicine, and plant chemistry allied to botany and agriculture, as sub-specialties.

Berzelius's university degree (Uppsala, 1802) was in medicine, and during 1804-1814 he wrote about the state of animal chemistry.[4] His first experimental efforts, however, dealt with the chemical analysis of mineral waters and minerals, the latter in association with Wilhelm Hisinger, a wealthy mine owner who greatly furthered Berzelius's career. Mineralogy continued to be one of Berzelius's main interests, and he later added several new metals to the list of those discovered by Swedish chemists. Largely self-taught in the antiphlogistic chemistry of his time, Berzelius eagerly accepted some of Lavoisier's ideas, notably those of acid radicals and the oxygen theory of acidity. He was also much attracted by the use of the Voltaic pile as an instrument in chemical research, and in 1803 he and Hisinger reported some experiments on the electrolysis of salts. Early in his self-education as a chemist, Berzelius became skilled in glass blowing, the assembly of laboratory apparatus, the preparation of chemical reagents, and the use of the blowpipe in the routine analysis of metals and minerals.[5]

Through his extensive writings, Berzelius played a significant role in the development of chemistry during the first decades of the nineteenth century.[6] The successive editions of his multi-volume textbook (German version, *Lehrbuch der Chemie*) were translated into other principal European languages, except English. There, Berzelius presented his theories and the results of his personal researches. As secretary of the Swedish Academy of Sciences, he also prepared an annual survey of the progress of chemistry; the form in which it was widely read and discussed was the German translation (*Jahres-Bericht über die Fortschritte der Chemie*). He conducted an active and wide-ranging correspondence with many European chemists, especially those in the German states. Also, he attracted to his laboratory some outstanding young German chemists, notably Eilhard Mitscherlich and Friedrich Wöhler; the latter became a loyal translator of the textbook and the *Jahres-Berichte*.

As mentioned previously, Berzelius was an early adherent of Dalton's atomic theory, and recognized the need for greater accuracy in the determination of combining weights than that achieved by Richter and later investigators. He began his systematic studies in this field in 1808 and ten years later presented analytical data for a large number of inorganic compounds, and a table of the calculated relative atomic weights (H = 1) for 45 elements.[7] He also introduced the use of a new symbolism for the designation of the composition of chemical compounds. Each element was denoted by means of the initial letter of its Latin or Greek name (in some cases a second letter was added, as in Cu for copper), with superscript numbers to indicate the calculated number of atoms of that element in the compound. This symbolism was gradually accepted with some modifications, notably the replacement of the superscripts by subscripts.[8] In using the experimentally determined ratio of combining proportions to calculate the numbers, Berzelius found Dalton's arbitrary rules, based on the axiom of simplicity, to be unsatisfactory. For example, Dalton considered water to be composed of one atom of hydrogen and one atom of oxygen, and from his finding that one part by weight of hydrogen combines with seven parts by weight of oxygen, he concluded that the atomic weights of hydrogen and oxygen must be in the ratio 1:7. By applying Gay-Lussac's law of combining gas volumes, Berzelius concluded that two hydrogen atoms combine with one atom of oxygen.

At first, Berzelius followed Dalton in using the principle of "greatest simplicity" in interpreting his extensive data on the composition of metallic oxides, and he wrote the formulas of ferrous and ferric oxides as FeO^2 and FeO^3 respectively, with double the value for the atomic weight of iron he later adopted.[9] He halved this value, and that for several other metals, in 1826 after Pierre Louis Dulong and Alexis Thérèse Petit showed in 1819 that the specific heats of the elements vary inversely as their atomic weights.[10] In the same year, Eilhard Mitscherlich discovered the phenomenon of isomorphism—the identity of the crystal form of chemical compounds of different, though related, composition—which provided an additional basis for the revision of some of Berzelius's first atomic weights.[11] The pioneer crystallographer, Haüy, who had declared that every crystalline form is characteristic of only one substance of definite composition, declined to accept Mitscherlich's claim, which was largely confirmed by later crystallographers.[12] Mitscherlich also described the crystallization of the same substance in two different forms (dimorphism), thus disproving Haüy's assertion that every substance of definite composition can occur in only one crystalline form, and later work provided other examples of such polymorphism. The finding of dimorphism served as the starting point for Louis Pasteur's famous studies on molecular dissymmetry, of which more later.

Nor were Berzelius's leading chemical contemporaries ready to accept his method of determining the number of each kind of atom in compounds. In 1832, Jean Baptiste Dumas wrote:

Berzelius in his treatise on chemical proportions, which marks so important an epoch in the history of the science . . . was the first to attack this difficult problem in its full scope. Without any rules to guide, he fixed by intuition the atomic weight of each substance, and usually allowed himself to be influenced by analogies which subsequent experience has only tended to confirm. But chemists have always wished that this arbitrary method, so successfully used by Berzelius, might be supplanted by something more fixed, more accessible to all kinds of intellect, and less subject to the capricious modifications of each writer.[13]

Many chemists of the time, including Justus Liebig, preferred to use "equivalents," directly given by the experimentally determined ratios of combining weights, introduced in 1814 by Wollaston, whose earlier adherence to Dalton's atomic theory had weakened.[14] This resulted in the halving of Berzelius's atomic weights, with C = 6, O = 8, S = 16, etc. and confusion in the chemical literature that lasted well into the 1850s. What brought the return of atomic weights after Cannizzaro's intervention in 1859 was the knowledge produced by organic chemical research about many substances to which the Avogadro-Ampère principle could be applied.

In parallel with his intensive effort to determine atomic weights, Berzelius developed a dualistic electrochemical theory, according to which

> . . . every chemical combination is wholly and solely dependent on two opposing forces, positive and negative electricity, and every chemical compound must be composed of two parts combined by the agency of their electrochemical reaction, since there is no third force. Hence it follows that every compound body, whatever the number of its constituents, can be divided into two parts, one of which is positively and the other negatively electrical.[15]

This theory, derived from his own electrochemical work and Davy's electrolysis of alkalis and alkaline earths, was based on the ancient concept of acid + base = salt, Lavoisier's theory of acidity (radical + oxygen = acid), and Berthollet's formulation of the mass action principle.[16] The theory stimulated much fruitful research and as its shortcomings became evident, Berzelius modified it, but he continued to defend its main features until the end of his life.

Berzelius set up a series of elements in which the most electronegative ones are oxygen and sulfur, which are followed by others of decreasing negative intensity, and then by still others of increasing electropositivity, with potassium as the most electropositive one.[17] He does not appear to have known that physicists considered the kind of unipolarity he assumed be a physical impossibility, and that Michael Faraday had made a distinction between the quantity and intensity of electricity.

For Berzelius, salts were combinations of oxides of opposite polarity; for example, he considered potassium sulfate to be composed of the binary electropositive (basic) KO and the binary electronegative (acidic) SO^3. In his scheme, salts that retain some polarity can combine with another binary unit, such as water. He also concluded that it is not oxygen but the radical that determines whether an oxide will be an acid or a base, and that the composition of salts such as potassium sulfate might be formulated as $(K)(SO^4)$. In 1832, this idea received decisive support in a paper by Wöhler and Liebig on the radical of benzoic acid.[18] This remarkable investigation, which represents a high point in nineteenth-century chemistry, began with a study of the benzoic acid found by earlier workers (notably Robiquet and Boutron-Charlard) to be formed upon oxidation of an oily material obtained from bitter almonds.

In this research, suggested by Wöhler and carried out in Liebig's laboratory, they corrected Berzelius's elementary analysis of benzoic acid and prepared a series of its derivatives. They concluded that the benzoyl radical acts as a unit, since treatment of the oil (benzaldehyde) with chlorine or bromine gave what they termed chlorbenzoyl or brombenzoyl, and the reaction of chlorbenzoyl with potassium iodide or ammonia gave iodobenzoyl or benzamid. In a letter to them dated 2 September 1832, Berzelius wrote:

> The results you have obtained from your investigation of the oil of bitter almonds are certainly the most important ones hitherto gained in vegetable chemistry, and promise to shed unexpected light on this part of science.

> The fact that a substance composed of carbon, hydrogen and oxygen combine with other substances, especially those which form salts and bases, establishes that there are ternary compound atoms (of the first order), and that the radical of benzoic acid is the first example to be demonstrated with certainty of a ternary substance which has the properties of a simple one.[19]

Apart from Berzelius's interpretation of the Wöhler-Liebig results in terms of his dualistic theory, their importance was greater than his praise allowed. Far from being significant only for plant chemistry, the discovery of the benzoyl radical spurred the search for other radicals and marked the beginning of the transformation of organic chemistry from animal and plant chemistry into the more general field of the chemistry of carbon compounds.

Berzelius included some organic compounds (for example, some of the plant acids isolated by Scheele) in his studies on combining proportions. For the carbon-hydrogen analyses, he modified the combustion apparatus of Gay-Lussac and Thenard (1811) so that the CO_2 was trapped by sodium hydroxide and the water by calcium chloride; Liebig's later, more rapid, method embodied the main features of this setup. In addition to his extensive separate writings on animal chemistry before 1814, and later in his textbook and the *Jahres-Berichte*, Berzelius made several personal research contributions in this field. In 1808 he isolated muscle lactic acid and found it to be different from the lactic acid obtained by Scheele from fermented milk. Later, he discovered pyruvic acid (1833), worked on cholic acid (1841), and ventured into plant chemistry in a study of chlorophyll (1833).

Although Berzelius stated in 1819 that "when organic atoms of the first order combine with inorganic atoms of the first order they follow (according to actual experiment) the same laws as compound inorganic atoms amongst themselves," [20] he also believed that:

> The structure of organic atoms is quite different from that of inorganic atoms. . . . Organic nature has its own particular way of producing oxides of compound radicals, and of giving to their chief constituents an electrical polarity that is quite independent of the different polarity that they have in inorganic nature. This is maintained in most cases only under the organic influence, but sooner or later all give it up to resume that which is proper to them in inorganic nature. Thence in all organic products spring the phenomena of destruction that we call fermentation and corruption, by means of which the elements gradually take up again their ordinary electrochemical characters. [21]

Over the years, Berzelius offered statements of similar nature on this subject. For example, in the 1839 *Jahres-Bericht* he wrote:

> In living nature, the elements appear to obey entirely different laws than in dead [nature]. . . . Seen from the viewpoint of a chemical investigator, the living body is a workplace in which there occur many chemical processes whose final result is to produce all the phenomena whose totality we term life. . . . After a slowdown in these processes, they finally cease, and from this instant the elements of the previously living body begin to obey the laws of inorganic nature.

> Every organic body therefore differs from an inorganic one in that the first has an observable beginning, it develops, ceases, and is destroyed, whereas on the contrary the inor-

ganic one was there before us, and continues to be stable, so that whatever the conditions it cannot be changed. The nature of the living body is therefore not based on its inorganic elements but on something [*Etwas*] different. . . . This something, which we term living force, lies entirely outside the inorganic elements, and is not one of their original properties, such as weight, impermeability, or electrical polarity, etc. . . . A force, incomprehensible to us, and foreign to dead nature, has produced this something . . . in a wonderful multiplicity, and designed with highest wisdom for definite purposes.[22]

Other statements of Berzelius's views suggest a more materialist outlook, and there has been some debate among historians about whether or not he was a vitalist.[23] Of particular interest is the recent discussion of Berzelius's attitude toward the flow of German romanticism into Sweden during the early nineteenth century.[24]

The issue of living force (*Lebenskraft*) came to the fore in 1828, with Wöhler's report of the formation of urea from ammonium cyanate (NH_4NCO).[25] This famous discovery was not a planned synthesis but an accidental event during Wöhler's systematic research on the chemistry of cyanates and evidence of his chemical acumen in recognizing the nature of the unexpected product. He did not claim that this discovery disproved the idea of a vital force or the impossibility of the artificial synthesis of animal substances. Some chemists considered it to be another case of isomerism, while others argued that urea was merely an excretory mammalian product and not a constituent of living matter. The relation of Wöhler's synthesis of urea to the fate of vitalism has attracted the interest of numerous twentieth-century scientists and historians.[26]

Some words about Wöhler's life and his chemical style. He acquired considerable chemical knowledge and skill as a *Gymnasium* student in Frankfurt; in 1821, when he began medical studies (first in Marburg, then in Heidelberg), he had already published two chemical papers, one on selenium, the other on cyanates. Wöhler continued the latter research in Heidelberg, where Leopold Gmelin was professor of chemistry. Wöhler intended to become an obstetrician, but Gmelin persuaded him to work with Berzelius. The year Wöhler spent in Sweden (1823-1824) was decisive in his career, for what began as a teacher-student relationship soon grew into a life-long affectionate friendship. In 1825, Wöhler became a teacher in the *Gewerbeschule* (technical school) in Berlin, where in addition to his work on cyanates, he made the first pure sample of aluminum, in 1827, and gave a description of its properties. In recognition of these achievements he was made a professor in 1828. Wöhler moved in 1831 to the *Gewerbeschule* in Kassel, but remained there only until 1836, when he was appointed professor of chemistry in the faculty of medicine at Göttingen, where he spent the rest of his professional life.[27]

As a scientist, Wöhler was above all a skillful and ingenious experimenter with a ready eye for interesting problems. He took from Berzelius his teacher's technical expertise and passion for analytical precision, but if Wöhler ever indulged in theoretical speculation, his numerous chemical papers show no sign of it. Like Berzelius, Wöhler had a special interest in the chemistry of metals and minerals, and most of his publications, especially after 1850, dealt with inorganic substances. Wöhler's medical training led him to study problems in physiological chemistry; for example, in 1838 he and Liebig published an important long paper (written by Wöhler) on uric acid.

Wöhler met Liebig in 1825, after Berzelius had settled a dispute in which Liebig wrongly criticized Wöhler's analysis of silver cyanate. Four years later, Wöhler wrote Liebig: "If you are so inclined, we could have fun to undertake some chemical work together, and make it known under both our names." [28] The collaboration began in 1830. During the succeeding 17 years they published about 18 joint papers, most of which (including the one on the benzoyl radical) were written by Wöhler. He collaborated with many other chemists, and was always scrupulous about the authorship of papers that dealt with his joint work. It was otherwise with Liebig. In 1840, Wöhler felt obliged to write his friend:

> If in citation of work that we have both done together only one of us is named, and especially in a journal in which both are named on the title page, about which everyone knows that you are the actual editor, and this editor allows that to happen and does not show the slightest consideration to report it, then everyone will conclude that this represents an agreement between us, and that the work is yours alone, and that I am a jackass.[29]

I return to Berzelius, and his role during the first half of the nineteenth century. Apart from his personal contributions in making stoichiometry one of the fundamental concepts of chemistry, he recognized in the discoveries of his time the emergence of other concepts to which he gave distinctive names. In addition to his quick appreciation of Mitscherlich's discovery of isomorphism, Berzelius was responsible for such lasting terms as isomerism, polymer, catalysis, and protein.

Before 1825 it was believed that substances composed of the same elements in the same proportions must have the same chemical properties. In that year, Liebig (working in Gay-Lussac's laboratory) found the same composition for the explosive silver fulminate as Wöhler had found in 1824 for the nonexplosive silver cyanate, and the repetition of their analyses gave the same results. Also, in 1825 Faraday reported that in olefiant gas (later named ethylene) the carbon and hydrogen were present in the same proportion as that in another more compressible hydrocarbon having double the specific gravity. In discussing these observations in 1832, Berzelius also cited Gay-Lussac's report that an acid isolated from tartar, and named racemic acid, had the same neutralization capacity as the tartaric acid isolated from Rochelle salt, but the two acids differed in solubility and crystalline form. Berzelius denoted these various pairs of substances, having similar composition and different properties, as "isomeric" and called Faraday's pair a case of "polymerism." [30] In succeeding years the number of examples of isomerism among organic compounds increased greatly, and the concept played a significant role in the development of ideas about the arrangement of atoms in molecules and in attempts at organic classification. [31]

In 1836, Berzelius wrote about alcoholic fermentation:

> . . . the conversion of sugar into carbonic acid and alcohol, as it occurs in the process of fermentation, cannot be explained by a double decomposition-like chemical reaction between a sugar and so-called ferment, as we name the insoluble substance under the influence of which the fermentation takes place. The substance may be replaced by fibrin, coagulated plant protein, cheese and similar materials, though the activities of these substances are at a lower level. However, of all the known reactions in the organic sphere, there is none to which the reaction bears a more striking resemblance than the decomposition of hydrogen peroxide under the influence of platinum, silver or fibrin, and it would be natural to suppose a similar action in the case of the ferment.

Berzelius ascribed this "similar action" to a new force:

> I do not consider this new force to be entirely independent of the electrochemical affinities of matter; I believe, on the contrary, that it is only a new manifestation of them, but so long as we cannot see their connection and mutual dependence, it will be more convenient to designate it by means of a separate name. I shall term this force, catalytic force.

I shall define catalysis as the decomposition of substances by this force, just as one defines analysis as the decomposition of substances by means of chemical affinity. . . . We have well-grounded reasons to conjecture . . . that in living plants and animals thousands of catalytic processes take place between the tissues and the fluids and result in the formation of the great number of different chemical substances, for whose production from the common raw material, plant juice or blood, no probable cause could be assigned. The cause will perhaps in the future be discovered in the catalytic force of the organic tissues of the organs of which the living body consist.[32]

In addition to the catalytic decomposition of hydrogen peroxide, observed by Thenard in about 1818, Berzelius cited the work of Döbereiner, mentioned above, and that of G.S.C. Kirchhoff in about 1812 on the conversion of starch to sugar by dilute sulfuric acid. Berzelius did not accept the claim of Schwann, Caignard-Latour, and Kützing that a living microorganism is responsible for alcoholic fermentation, a view strongly expounded by Pasteur around 1860. The dispute continued until the end of the nineteenth century, when Buchner prepared a cell-free yeast extract.[33]

The word "protein" first appeared in print in 1838, in a paper by Gerrit Jan Mulder, a Dutch chemist and a former student of Berzelius. In a letter to Mulder dated 10 July 1838, Berzelius wrote:

I consider it sufficiently well established that the immediate organic substances either are oxides of compound radicals or are combinations of two or even several oxides of this kind. It is necessary first to look for this radical. This is easy when it contains only two elements. The addition of nitrogen complicates matters a bit, but in general the difficulty is not great. . . . Now I presume that the organic oxide which is the base of fibrin and albumin (and to which it is necessary to give a particular name, *e.g., protein*) is composed of

a ternary radical combined with oxygen. . . . The word protein that I propose to you for the organic oxide of fibrin and
albumin, I would wish to derive from πρώτειοσ, because it
appears to be the primitive or principal substance of animal
nutrition that plants prepare for the herbivores, and which
the latter furnish to the carnivores.[34]

From his analyses of egg albumin, serum albumin, and fibrin
Mulder concluded that all three were composed of the same
unit (which he called protein), and that in egg albumin and
in fibrin it was combined with one atom of sulfur and one of
phosphorus, while serum albumin had two atoms of sulfur
and one of phosphorus. He claimed that treatment with
dilute alkali removed the sulfur and phosphorus and that
he had isolated the fundamental radical. He also claimed
that treatment of plant gluten with alkali gave a material
whose composition was the same as that of the unit obtained
from the animal sources. These conclusions were accepted
by Liebig until 1846 when his student, Nicholas Laskowski,
showed that Mulder's supposedly sulfur-free protein contained sulfur. The word protein fell into disrepute, and terms
like "albuminous body" were generally used, and it did not
come back into favor until the turn of the century, when it
was adopted by Emil Fischer.[35]

The dualistic radical theory, initially widely accepted
and later stubbornly defended by Berzelius, stimulated new
approaches to the problems raised by a growing number
of well-defined organic compounds and their reactions.
Apart from those who rejected the atomic theory, some
chemists, particularly in France, began to resist dualism.
In 1824 Chevreul, whose work on the constitution of fats
represented the first clear-cut study of a set of complex
organic compounds, expressed doubts. The ideas of Ampère
attracted the attention of Marc Gaudin and of Alexandre
Edouard Baudrimont. The latter wrote:

> No sort of combination which contains more than two elements should be represented by a bi-binary formula, whether these combinations be organic or inorganic in origin. In all combinations every atom plays a significant part; hence we should consider it as inexact not only the Guytonian nomenclature for bodies containing three elements or more, but all classifications based on this nomenclature, and finally, Berzelius' electrochemical theory.[36]

The ideas of Gaudin and Baudrimont were welcomed by crystallographically minded chemists, notably Auguste Laurent (of whom more shortly), but were dismissed by Jean Baptiste Dumas, who did not reject electrochemical dualism until 1839. In his *Jahres-Berichte*, Berzelius ridiculed these views, as he did consistently in later years in writing about Dumas, Laurent, and Gerhardt.

As mentioned earlier, the Wöhler-Liebig studies on the benzoyl radical spurred efforts to identify other organic radicals. In 1833, Robert John Kane suggested that C_2H_5 (C = 12) is present in alcohol and ether; Liebig named it ethyl. The methyl (CH_3) radical, so named by Berzelius, was discovered in 1834 by Dumas and Eugène Peligot in a brilliant investigation of the long-known wood-spirit. They also showed in 1836 that a compound discovered by Chevreul is cetyl alcohol, containing the cetene radical ($C_{16}H_{34}$). With Polydore Boullay, Dumas had previously proposed that alcohol and ether belong to a series derived from a two-carbon unit (Berzelius named it etherin), and Liebig added the acetyl radical (C_4H_6; now C_2H_3O) to the series. Liebig's organic chemical contributions also included the recognition of the importance of Thomas Graham's studies in 1833 on the four known phosphoric acids. On the basis of work done in 1836 by his Parisian friend Theophile Jules Pelouze on organic polyacids such as citric acid, Liebig concluded that they are combinations of an oxygen-containing radical with hydrogen atoms, rather than as formulated by Graham (and Berzelius) as radical + H_2O. The ensuing argument between Liebig and Berzelius on this subject put a strain on their previously cordial relations. A complete break came in 1842, when Berzelius criticized Liebig's book on animal chemistry.

Among the organic radicals discovered during the 1830s a special place is occupied by the one Berzelius named "cacodyl." It was discovered by Robert Bunsen, whose academic career began at Kassel as Wöhler's successor, and continued as professor at Marburg (1839-1851), Breslau (1851-1852), and Heidelberg (1852-1882).[37] At Kassel, Bunsen undertook the study of organic compounds of arsenic, many of which are highly toxic, foul-smelling, and explosive. In this research, both courage and exceptional experimental skill and ingenuity were required; the latter qualities were evident in all of Bunsen's later scientific work. He isolated and characterized a substance ($C_4H_{12}As_2$), which he considered to be a radical, and prepared several derivatives. This line of investigation on metallo-organic compounds was continued by his students Edward Frankland and Hermann Kolbe (of whom more later) and by August Cahours.

Bunsen's work on the cacodyl series was his only major organic chemical effort. Although perhaps best known for his invention of the Bunsen burner (it replaced the blowpipe in mineralogical analysis), his later contributions included the development of new methods of gas analysis, improvement of the galvanic battery, and the use of electrochemical techniques for the preparation of pure metals, photochemical studies (with Henry Roscoe) and, above all, the introduction, with Gustav Kirchhoff, of spectrochemical analysis.[38] In contrast to the leading organic chemists of his time— Berzelius, Liebig, Dumas—Bunsen had an excellent knowledge of mathematics and physics.

At Marburg, and later at Heidelberg, Bunsen's lectures

were much admired by his students. John Tyndall, who was at Marburg in 1848, wrote:

> Bunsen was a man of fine presence, tall, handsome, courteous, and without a trace of affectation or pedantry. He merged himself in his subject; his exposition was lucid, and his language pure; he spoke with the clear Hannoverian accent which is so pleasant to English ears; he was every inch a gentleman. After some experience of my own, I still look back on Bunsen as the nearest approach to my ideal of a university teacher.[39]

Bunsen and Liebig were men of entirely different make-up. This was appreciated by Liebig's own students; in an obituary notice for Friedrich Knapp (a pupil and brother-in-law of Liebig) it is stated:

> There was a lively traffic of the younger Giessen chemists to Marburg; they traveled as often as possible to visit the much admired Bunsen. As Knapp reported, "Liebig and Bunsen did not understand each other particularly well, although there was of course mutual respect. Each went his own way, and Bunsen was a thoroughly original character, he was no-one's pupil, and therefore not likely to be attracted to Liebig. But we younger men had an immense admiration for Bunsen. He spoke beautifully, so that just to hear his voice gave me pleasure. At first he was not communicative, rather reticent; but as acquaintance grew, he became intimate."[40]

This comparison invites some words about Liebig.

Unquestionably, Liebig was the best-known nineteenth-century chemist. He has also been called "the greatest chemist of his time."[41] Certainly, if greatness is measured in terms of a scientist's influence on the development of reliable knowledge, Liebig's research contributions justify his inclusion among the great chemists of the century, but the superlative seems excessive. Liebig's public renown did not come in response to his organic-chemical research, but rather came from his activity as an entrepreneur and propagandist for the place of chemistry in universities,[42] and above all, for the importance of his kind of chemistry in agriculture and the medical sciences, especially physiology.[43]

Liebig obtained his doctorate in 1823 from Erlangen by somewhat devious means, and received a stipend to continue his chemistry studies in Paris. There, he won the interest of Gay-Lussac in his work on fulminates, begun in Erlangen, and also attracted the attention of Alexander von Humboldt, always on the lookout for promising young scientists. Humboldt wrote Grand Duke Ludwig of Hessen-Darmstadt recommending Liebig for a professorship, with the result that in 1824 he became associate professor of chemistry at Giessen, and a full professor the following year. Inspired by the vibrant scientific life in Paris, Liebig decided to be not only a teacher but a research chemist who taught future research chemists.

Liebig did not achieve that aim until about 1838, when he began to grow tired of teaching and became more interested in writing about agriculture and physiology. Before then, the great majority of his students had matriculated in pharmacy, and there were at most ten laboratory students. In this respect, therefore, he had merely added another place where future pharmacists could get their chemical training.[44] The attractiveness of Giessen for such students was that, in contrast to major German universities, matriculants were not required to have completed a full *Gymnasium* program, nor to have passed the *Abitur* (a leaving examination). Between 1838 and 1852, when he left for Munich, many future professors of chemistry spent some time with Liebig.[45] Toward the end of this period, however, he appears to have paid less attention to promising students. Thus, he failed to recognize the talent of a young architecture student, August Kekulé, who had been inspired by Liebig's lectures to switch to chemistry. There can be no doubt that Liebig's teaching enterprise in Giessen, with an emphasis on the laboratory training of able students in the preparation of pure substances and in their elementary analysis, and to which he gave much publicity in his *Annalen*, promoted the expansion of chemical laboratory instruction at German universities. It should also be noted that after about 1850 the demand for the chemical training of pharmacists was exceeded by that for well-educated chemists for the burgeoning German chemical industry.[46]

Liebig's style as a productive organic chemist, especially in his joint research with Wöhler, changed during the 1840s into that of a propagandist for the role of chemistry in agriculture, physiology, and pathology, fields in which he had no technical experience. In several of his books [47] he developed a theme based on a Newtonian view of matter in motion reminiscent of the ideas of Willis and Stahl, to which he added the once fashionable concept of *Lebenskraft* (vital force) being responsible for the chemical transformations (*Stoffwechsel*) in living organisms:

> I will now direct the attention of scientists to a previously unnoticed cause which brings about metamorphoses and decomposition phenomena which are usually called decay, putrefaction, fermentation, and mouldering. This cause is the ability of a substance engaged in decomposition or combination, i.e. in chemical action, to give rise in a substance in contact with it the same ability to undergo the same change which it experiences itself. [48]

Liebig rejected Berzelius's idea of a catalytic force, and explained alcoholic fermentation by proposing that oxygen causes the decomposition of the gluten of yeast and that this motion is communicated to the sugar undergoing fermentation.[49] With his customary stinginess in citing those whose ideas resembled or foreshadowed the ones he claimed as his own, Liebig did not note that in 1834, Mitscherlich wrote that "decompositions and combinations of this kind are very frequent; we will call them decompositions and combinations by contact."[50] In 1842, Liebig advised Mitscherlich "not to bother with his old wives' chatter [*Altweiberschwatz*], and that he stop seizing upon results of investigations he did not perform."[51]

In his essay on the discovery of energy conservation, Thomas Kuhn included Liebig among those who, during 1837-1844, expressed views about the convertibility of forces.[52] Another was Liebig's friend Carl Friedrich Mohr, who proposed in 1837 that heat is a form of motion. It is difficult to imagine that Liebig did not know about Mohr's ideas, or about Robert Mayer's 1842 paper in Liebig's *Annalen*. Moreover, Liebig's concept of the vital force bears some resemblance to that of Berzelius, and somewhat more to that of the physiologist Johannes Müller, who in 1837 wrote about the substances excreted by animals:

> As those excretions are constant, even when the supply of nutrient is stopped, it necessarily follows that a constant decomposition of the substance of the body is essentially connected with life. It cannot, indeed, be otherwise if it be true, as it has already been proved to be, that the vital force is manifested in an animal body only when certain vital stimuli produce in the living tissues constant material changes, of which the phenomena of life are merely the external signs, just as the flame is the appearance resulting from material changes effected in combustion.[53]

In his excursions into pathology, Liebig asserted that contagion was not caused by living agents, but by the formation in the blood of toxic nitrogenous degradation products or the ingestion of such products of putrefaction and decay. These views received considerable attention from British physicians and promoted efforts to improve sanitation,[54] but also brought him into collision with the redoubtable Louis Pasteur, who outmatched Liebig in their scientific debate.[55]

The branch of physiology in which Liebig had the greatest influence was human nutrition. He adopted (without attribution) William Prout's classification of foodstuffs as saccharine, oily, and albuminous, and divided them into two groups: the "respiratory" portion of the diet included the non-nitrogenous fats, starch, sugar, etc., which Liebig considered to be the fuel of the body; and the "plastic" elements represented by plant proteins (eaten by herbivorous animals), and animal flesh and blood (eaten by carnivores), which were converted into blood. In his view, the central place in the machinery of the animal body was occupied by the vital proteins of the blood and the tissues, especially muscle. During muscular exercise or starvation, the respiratory oxygen attacks these vital proteins, with the formation of uric acid, urea, and ammonia, which are excreted in the urine. Comparison of the elementary composition of uric acid and urea led Liebig to conclude that uric acid is a metabolic intermediate between protein and urea. He also used the data from the elementary analysis of ingested foods and excreted end products (including exhaled carbon dioxide), as well as respiratory oxygen, to follow Lavoisier in setting up a balance sheet based on a chemical equation. In Liebig's case, the equation was based on the assumption that urinary urea nitrogen is solely derived from protein nitrogen. This intake-output approach to the study of metabolism was developed in modified form by later German physiologists, notably Max Rubner, and in the United States by Wilbur Olin Atwater.[56] Combined with the thermochemical approach introduced by Lavoisier and Laplace, and developed during the nineteenth century by Dulong and Despretz, Hess, Thomsen, and Berthelot, the analytical determination of the elementary composition of dietary materials and excreted products, together with their heats of combustion, became known as "energy metabolism." [57] For the physiologist Claude Bernard, however, the intake-output method was misguided because it told nothing about the intermediate steps in the conversion of foodstuffs into excretory products.[58]

Few famous books in the history of physiology approach the Popperian ideal of scientific theory more closely than the first edition of Liebig's *Animal Chemistry* in the number of speculations that were refuted experimentally. The best-known example is Liebig's theory that the oxidative decomposition of protein is the cause of muscular contraction, and its refutation by Adolf Fick and Johannes Wislicenus in their celebrated mountain climb.[59] It should be noted, however, that Liebig's conclusion from the similar elementary composition of starch and fats that herbivorous animals can make fat from carbohydrate was heatedly disputed until 1844, when Jean François Persoz proved Liebig right during the course of feeding geese an unrestricted diet in order to produce maximum amounts of fois gras.

Liebig's chemical physiology may have been questioned by eminent physiologists of his time,[60] but the public renown gained from the sale of "Liebig's Meat Extract" was not diminished by it. In a book published in 1847, he described the preparation and nutritive value of this concoction, which contained, among other ingredients, lactic acid, creatine (discovered by Chevreul in 1832), and two other nitrogenous substances discovered by Liebig—creatinine (derived from creatine by the loss of H_2O) and inosinic acid (which also contained phosphorus). After he moved to Munich in 1852, Liebig encouraged the large-scale commercial production of his meat extract, and the inclusion of his name in the publicity attending its sale.[61] Liebig also made the investigation of various foods a major research activity, and in 1866 he devised a formula for infants [Suppe für Säuglinge]. He also considered it appropriate to bring these contributions to the attention of his chemical colleagues in the pages of his *Annalen der Chemie.*

In the first edition (1840) of his *Agricultural Chemistry* (there were to be seven more during his lifetime) and his other early writings on this subject, Liebig questioned the wide use of humus (a topsoil product of plant decay) as a source of carbon, on the ground that atmospheric CO_2 was sufficient for this purpose. He asserted that the plant nitrogen came from ammonia in rainwater, a view that was disputed by Jean Baptiste Boussingault and Georges Ville in France and by John Bennet Lawes and Joseph Henry Gilbert in England.[62] It was free nitrogen that was taken up from the atmosphere. Later work showed that nitrogen is "fixed" by symbiotic organisms in the root nodules of legumes and converted to combined nitrogen.[63]

Liebig rightly emphasized the importance of replenishing the minerals in the soil,[64] and he devised a series of artificial fertilizers containing alkali and phosphate, which were patented and manufactured in England by his friend James Muspratt and his sons. When tested by Lawes at his Rothamsted experiment station, most of these fertilizers proved to be ineffective. Lawes entered the market by developing an artificial fertilizer containing "superphosphate," obtained from excavated human bones. Liebig's reaction was expressed in a letter to Faraday, dated 27 July 1856: "Mr. Lawes is, I believe, a manufacturer of manure, and by my disputing his scientific position and showing that his conclusions are erroneous he thinks to lose his customers."[65]

I return to the state of organic chemistry in the 1830s and the fate of the radical theory of Lavoisier and Berzelius. In 1837, during one of their periods of amity, Liebig and Dumas stated in a joint paper (written by Dumas) that "in inorganic chemistry the radicals are simple; in organic chemistry they are compound—that is the only difference." [66] By that time, however, work by Dumas and other chemists (notably Auguste Laurent, Faustino Malaguti, and Henri Victor Regnault) on the chlorination of organic compounds had raised questions about the dualistic radical theory. No less inclined than Liebig to chemical speculation, Dumas had proposed in 1834 his "substitution" theory, which stated that chlorine atoms can replace hydrogen atoms. But he still retained his adherence to the electrochemical theory, while Liebig offered a more restrictive definition of organic radicals.[67] In 1839 Dumas reported that the trichloroacetic acid he had prepared by the chlorination of acetic acid in sunlight did not differ appreciably in its chemical properties from acetic acid. He proposed a theory of "types," which are conserved when such substitution of hydrogen by chlorine is effected, and also rejected the electrochemical theory.[68] Jules Pelouze, Liebig's chief ally in Paris, wrote to Berzelius that he was convinced that Dumas was using the theory of substitutions as a device to elevate himself to the position of *Chef de l'Ecole* of the new organic chemistry,[69] and Bunsen stated that his results on the cacodyl series "contradicted the opinions that the new French chemical school has tried to introduce into science . . . this radical acts in complete opposition to the premises of Dumas's theory of substitution." [70]

Some features of the life and style of Jean Baptiste Dumas resembled Liebig's. Like Liebig, Dumas entered chemistry after serving as a pharmacist's apprentice, but instead of going to Paris he went to Geneva. He spent six years (1816-1822) there working with Augustin Le Royer on several chemical problems, and also did physiological research with Jean Louis Prevost. Like Liebig, Dumas attracted the attention of Alexander von Humboldt, who urged him to go to Paris and wrote about him to friends there. During 1823-1832, Dumas applied the Avogadro-Ampère theory in an attempt to use vapor pressure measurements to determine atomic weights. He held several teaching appointments in Paris, among them professorships at the Ecole Polytechnique and the medical school, and his rise to prominence was certified in 1832 by his election to the Académie des Sciences. Apart from demonstrating his ability as a lecturer and a research chemist, Dumas showed his talent as an entrepreneur, for example, as one of the founders of the Ecole Centrale des Arts et Manufactures. Also, in 1826 he married the daughter of the noted mineralogist Alexandre Brongniart, and acquired as a brother-in-law Adolphe Brongniart, who subsequently became inspector general of higher education in the sciences.

Like Liebig, during the late 1830s Dumas became actively interested in agricultural chemistry, and collaborated with Boussingault in studies on the nitrogen metabolism of plants, using a new method Dumas had devised for the nitrogen analysis of organic compounds. In 1841 he and Boussingault published their *Essai du Statique Chimique des Êtres Organisées*, in which they drew a sharp distinction between the physiology of plants and animals.[71] In particular, Dumas (with Auguste Cahours) questioned the conclusions of Mulder and Liebig during the early 1840s on the identity of plant and animal proteins.[72] There ensued a series of polemical exchanges, largely on matters of priority and analytical accuracy.

After 1848, Dumas entered French politics, and became Minister of Agriculture and Commerce (1849-1852) and a senator. A cartoon by Honoré Daumier depicted Dumas dressed in a laboratory apron, with a furnace and alembic in the background. The inscription read: "A new chemical prodigy—Dumas has succeeded in getting out of his retort a portfolio! Since being in the ministry, Dumas has always taken care to avoid the platform, he gives as an excuse that he is always busy analyzing the speeches of the other orators." Dumas continued to be a professor at the Sorbonne until he retired in 1867, and one of his chemical interests was the arithmetical relationships among the equivalents for similar elements, a popular subject that had also been taken up by (among others) William Odling, Adolf Strecker, J.A.R. Newlands, and Dmitri Mendeleev.[73]

Between 1828 and 1848, Dumas had about 39 research students, six of them from abroad.[74] Those from France included Cahours, Gerhardt, Laurent, Malaguti, Pasteur, Sainte-Claire Deville, and Wurtz. Marignac (Switzerland), Melsens and Stas (Belgium), and Piria (Italy) were among the foreign students. There were none from Britain or Germany. Two of these men—Laurent and Gerhardt—occupy a central place in the mid-nineteenth century development of organic chemistry.

Auguste Laurent, the older of the two, came to Paris in 1826 to study mineralogy and crystallography at the School of Mines, and shortly after receiving his degree he became Dumas's research assistant (1830-1831). He then worked at the Sèvres porcelain factory, headed by Dumas's father-in-law. There, Laurent made a significant contribution to the crystallography of silicates, but since his principal duty was to perform routine analyses, he resigned in 1834 to pursue an independent career. During the next four years he did research in a laboratory provided by a perfumer and then in a private laboratory. In 1838 Laurent joined a ceramics factory in Luxembourg, then was professor of chemistry at Bordeaux during 1839-1845, and returned to Paris in the hope of getting a professorship there. He obtained a laboratory at the Ecole Normale, where Dumas's student, Pasteur, had begun his research, and then at the Mint, where Laurent set up his laboratory in a damp basement. In 1850, he applied for the professorship at the Collège de France, to be vacated by

Pelouze, and was bitterly disappointed when the Académie des Sciences overrode the favorable vote of the professors, as well as the strong recommendation of the noted crystallographer Biot, and rejected his application. Laurent felt, perhaps rightly, that Dumas could have intervened more actively in his behalf, but it is also likely that Laurent's staunch republicanism disturbed the academicians at a time of political uncertainty. Shortly afterward, Laurent fell ill and died of tuberculosis in 1853. In his final months, he composed his *Méthode de Chimie*, published a year later. William Odling prepared an English translation (1855), with some editorial corrections, and it was in this form that Laurent's book became famous.[75]

Laurent's talent as a skilled experimenter and an imaginative theoretician became evident in his first organic chemical investigation, suggested by Dumas, on naphthalene, a substance known to be composed of only carbon and hydrogen. After working out a method for its purification from coal tar, during the course of which Laurent also discovered anthracene, he proceeded to study systematically the reactions of naphthalene and its derivatives with chlorine and other reagents. Among his findings was the discovery that treatment of chloronaphthalene with nitric acid produced nitronaphthalene with the release of HCl. By 1836, his accumulated data led him to conclude that "all organic compounds are derived from a hydrocarbon, a fundamental radical, which often does not exist in its compounds but which may be represented by a derived radical containing the same number of equivalents."[76] Laurent was greatly influenced by Baudrimont's rejection of the dualistic electrochemical theory, as well as by two earlier theories proposed by Dumas — the two-carbon (etherin) theory of the composition of alcohol and ether, and the substitution theory. Laurent's training in crystallography also led him to apply Haüy's concept that the fundamental structure of a crystal corresponds to the unitary molecular structure of its parts. To this analogy, Laurent added the principle of isomorphism, and he assumed that in organic compounds it is not the nature of the atoms in a molecule that determine its properties but their position and order. He drew geometric models, including symmetrical prisms in which the corners are occupied by carbon atoms and the midpoints of the edges by hydrogen atoms. He considered such a prism to serve as a fundamental "nucleus" for substitution reactions. By means of such models Laurent explained the substitution of a hydrogen atom by a chlorine atom as the entry of the halogen into what he called the "nucleus" without a change in the properties of the molecule, predicted that this would apply to the chlorination of other compounds including acetic acid,[77] and also used these models to propose a rational basis for the classification of organic compounds.[78] In 1838, Laurent's theory, which involved the replacement of an electropositive hydrogen by an electronegative chlorine, drew the fire of Berzelius and was disavowed by Dumas:

> Berzelius attributes to me an opinion precisely contrary to that which I have always maintained, namely that chlorine takes THE PLACE of the hydrogen. . . . The law of substitution is an empirical fact and nothing more; it expresses a relation between the hydrogen expelled and the chlorine retained. . . . To represent me as saying that hydrogen is replaced by chlorine, which fulfills the same functions, is to attribute to me an opinion against which I protest most strongly, as it contradicts everything that I have written on this subject. I am not responsible for the gross exaggeration with which Laurent has invested my theory; his analyses moreover do not merit any confidence.[79]

In his detailed reports of 1839 and 1840, mentioned above, on the properties of the trichloroacetic acid he had prepared by the chlorination of acetic acid, Dumas executed a volte-face, announced that the two compounds had similar properties, abandoned the electrochemical theory, and proposed a "type" theory. There then ensued a priority dispute between a resentful Laurent and an ungenerous Dumas, who wrote in 1840 that "Laurent stressed the identity of the role of chlorine and of hydrogen in substances formed by substitution, long before experiment had decided positively on this matter," [80] and only later (after Laurent's death) did Dumas state: "That which Laurent later recognized is that in the substitution phenomena the type is conserved, that is to say that not only does the chlorine take the place of hydrogen, but also plays the same role." [81] It should be noted that in 1838 Liebig led the chorus denigrating Laurent's theory.[82]

During the last decade of his life, Laurent modified his theory to account for the result of his experiments on the bromination of camphor, and offered additional evidence in its favor, notably in a masterly study of the derivatives of isatin, which he had discovered as a product of the oxidation of indigo.[83] This work was praised highly by Berzelius, who compared it to that of Wöhler and Liebig on the benzoyl radical.

In 1843, Laurent met Charles Gerhardt. They became close friends and, two years later, fellow outcasts. At that time, Gerhardt was chargé des cours at Montpellier; he was made a professor there in 1844. Like Laurent, Gerhardt was a republican. An Alsatian, fluent in French and German, he began his chemical studies in Karlsruhe and Leipzig, and decided on a career in chemical research rather than satisfying his father's wish that he work in the family factory. Deprived of his father's financial help, Gerhardt impulsively entered military service but won his release soon afterward, and spent several months (1836-1837) in Liebig's laboratory. This association provided some income from the translation of a number of Liebig's books into French. Liebig urged Gerhardt to go to Paris, where he completed his chemical studies, was Dumas's teaching assistant, and received his doctorate in 1841. Gerhardt was then shipped off to Montpellier, where the inadequate facilities for chemical research (a common feature at most provincial French universities) and the hostility of his colleagues made him very unhappy. By 1846, his letters to Laurent became increasingly desperate, and Laurent's efforts in Gerhardt's behalf, at a time when his own situation in Paris was difficult, say much about Laurent's generosity of spirit.[84] Gerhardt's application for a leave of absence from Montpellier was denied, and in 1851 he resigned, joined Laurent in his laboratory at the School of Mines, and set up a private chemistry school. Gerhardt also undertook the preparation of a multi-volume chemistry textbook, which appeared during 1853-1856. In 1855, he was appointed professor of chemistry and pharmacy at Strasbourg (his birthplace), probably because Dumas chose not to oppose it. There, he found a well-equipped laboratory, and could look forward to many years of productive activity, but he died a year later after an attack of acute peritonitis.

Although their names have been linked inseparably by historians, Laurent and Gerhardt had different scientific preconceptions and different styles of exposition.[85] They disagreed about the value they attached to the study of the physical properties of chemical compounds. Although Gerhardt used the regularities found by Hermann Kopp in the boiling points of related compounds, he did not accept Laurent's emphasis on the importance of crystallography and optical activity. They differed when, in 1845, Gerhardt used intemperate language to criticize Liebig's work on "mellon" (a new radical that Liebig formulated as CNS), and drew from Liebig a vitriolic response.[86] In a letter to Laurent, for whom he had greater respect, Liebig wrote: ". . . your idea of associating with M. Gerhardt was the worst misfortune that could befall you. . . . If you support him you alone will be the loser, for he has nothing to lose. Read his article carefully and tell me whether that man has truth in his soul."[87] In their correspondence, Laurent urged Gerhardt to stick to the facts of the case, and to omit passages that infuriated Liebig. After further blasts from Liebig, in which he included Laurent, the latter wrote to Liebig: "It is you, covered with honors, replete with wealth, it is you, I say, who lowers yourself for the third time to play toward me the role of a base slanderer."[88] In 1850, Gerhardt sought to make his peace with Liebig, who responded favorably. Indeed, in 1852, when Gerhardt published his most important experimental paper—on the first preparation of acid anhydrides—Liebig praised it highly. In one of his papers on acid anhydrides, Gerhardt presented a revision of Dumas's theory of types, of which more in the next section.[89]

In 1838 and 1839, while still a student, Gerhardt published two theoretical papers in which he challenged the ideas of Berzelius, Liebig, Dumas, and Laurent about radicals, and asserted that all organic compounds are derivatives of hydrocarbons in which various substituents had replaced the hydrogens.[90] These papers were briefly dismissed by Berzelius, and Gerhardt turned to experimental work. On the advice of Liebig that he should attach himself to Dumas, Gerhardt wrote in one of his papers: "These facts are a new proof in favor of two fundamental principles put forward by M. Dumas—they are entirely in conformity with substitutions, and demonstrate the constancy of chemical properties in derived substances of a type by substitution."[91] Soon after his appointment at Montpellier, which he owed to Dumas, Gerhardt returned to the problem of organic classification, and developed the concept of homologous series; for example, he considered organic acids to form a series whose members differed only in the number of CH_2 groups. In their correspondence, Laurent criticized his friend's approach, and called attention to his neglect of isomerism.[92] In 1842, Dumas wrote that several properties of fatty acids bore a relation to the number of their carbon atoms,[93] and later claimed priority for Gerhardt's idea of homology, which Laurent had criticized in their correspondence. Gerhardt did not heed Laurent's plea to desist from pursuing the matter, and thus earned Dumas's enmity not only toward Gerhardt but also, through guilt by association, toward Laurent as well.

Of greater historical interest is the interaction of Laurent and Gerhardt in the discussion of the meaning of the terms atom, molecule, and equivalent in relation to the application of the Avogadro-Ampère principle to vapor density measurements. In 1843, Gerhardt stated that "atoms, equivalents, and volumes are synonymous . . . the densities of gases are proportional to their equivalents."[94] He also returned to the "two-volume" convention, and adopted the atomic weights of Berzelius for some metals. On the other hand, in 1841 Laurent had made a distinction between atoms and equivalents and, in 1846, after the publications of Avogadro on gas volumes and of Regnault on atomic heats, had distinguished between atoms, molecules, and equivalents.[95] Laurent defined an equivalent as the amount of an element that must be used to replace another element and play the same role in reactions of a similar type.[96] Gerhardt accepted this definition in 1849, and he and Laurent defined a molecule as the smallest amount of a substance that participates in a chemical reaction, and an atom as the smallest amount of the individual elements that compose the molecule. The word molecule was intended to replace the term "compound atom," which Berzelius also applied to radicals having more than one element; Gerhardt renamed such parts of a molecule "residues."

In his famous *Sketch of a Course of Chemical Philosophy*, distributed by his friend Angelo Pavesi at the 1860 Karlsruhe Congress, Stanislao Cannizzaro presented a convincing argument for the applicability of the Avogadro-Ampère principle to the determination of relative atomic weights from measurements of vapor densities and specific heats, as well as the stoichiometry of chemical reactions. In his account of previous work, he mentioned Dumas's vapor density studies (1826) and also Gerhardt, whom he criticized for not having fully extended his approach. The name of Laurent appears only once, in conjunction with Gerhardt. Another interesting feature of Cannizzaro's treatment is his usage of Berzelius's terms electropositive and electronegative radicals, with which most chemists of the time were more familiar than they were with the chemical language of Laurent. Cannizzaro's definition of equivalents is much like that of Laurent, mentioned above.[97] Cannizzaro is also remembered for his 1853 discovery of a reaction that still bears his name—the disproportionation of an aldehyde by KOH to the corresponding acid and alcohol (in his case, benzaldehyde to benzoic acid and benzyl alcohol).

Chapter Six

VALENCE AND MOLECULAR STRUCTURE

༂

IF THE ORGANIC chemistry of the 1830s was dominated by the theory of radicals, with the benzoyl radical of Wöhler and Liebig and Bunsen's cacodyl radical in the foreground, the organic chemistry of the 1840s was increasingly dominated by the theory of types. There had been foreshadowings in earlier writings; for example, in 1840 Liebig suggested that ammonia might be regarded as "der Typus aller organischen Basen."[1] The most imaginative proposal of the 1830s embodying the theory of types was the "nucleus" theory of Laurent, mentioned above. From his studies of substitution reactions, and analogy to the structure of crystals, Laurent sought to devise a classification of organic compounds based on rational formulas. In his later definitive paper on this subject, he wrote:

> I do not have the pretension of representing by my formulas the real arrangement of the atoms, according to M. Chevraul's expression, the *absolute* formulas. It is indeed ideas about the molecular arrangement which have guided me in the system I have proposed; but one may, if one wishes, to abstract from them and not to see anything but symbols whose features recall at a glance not only the composition and the nature of the substance which they represent, but also the series to which the substance belongs and the place it occupies in this series.[2]

The novel terminology offered by Laurent gained little favor. Instead, the one later used by Gerhardt, and based on the theory of types presented in 1840 by Dumas, served as a basis for further discussion.

In Dumas's theory, what he called his "law of substitution" was extended in his proposal that "in an organic compound, all the elements can therefore be successively displaced and replaced by others."[3] According to Pelouze, the theory was privately derided by Dumas's academic adversaries in Paris,[4] and there also appeared in Liebig's *Annalen der Chemie* a satirical paper (written by Wöhler), which reported that all the atoms in manganese acetate had been successively replaced by chlorine atoms to yield a product that had retained the properties of manganese acetate.[5]

In his important experimental paper on acid anhydrides, Gerhardt reformulated the theory of types in terms of groups of atoms (he called them "residues'), revived the electrochemical idea of positively and negatively charged groups, and considered organic compounds to be derived from H^2O, NH^3, HCl and H^2.[6] By that time, Adolphe Wurtz and August Wilhelm Hofmann had provided strong experimental evidence for the existence of the ammonia type of organic compounds, Alexander Williamson had done the same for the water type, and Edward Frankland's work on zinc ethyl had enlarged the scope of the theory of types.[7] More will be said about these four chemists shortly. At this point, I call attention to Laurent's admission of the kind of error one can make in applying one of the theories of types:

> I placed aniline by the side of phenic acid and benzine, although my formula $C^6H^5.H^2N$ indicated, that it might equally well be placed by the side of ammonia. . . . Seeing that aniline remained basic like the ammonia from it sprung, I ought, in accordance with the principles I had enounced with reference to substitutions, to have regarded it as an ammo-

nia with a phenic residue. This last mode of considering the subject is the one adopted by Hofmann, who, viewing it in a particular aspect, has drawn from it one of the most ingenious ideas ever promulgated in chemical science. He has represented the phenic residue C^6H^5 as a radical analogous to ethyl, or a metal, and, consequently, capable of replacing an atom of an atom of hydrogen, without affecting the basic properties of the ammonia into which it enters.[8]

In his book, Laurent also took occasion to criticize Berzelius's use of Gerhardt's concept of the "copula" (an organic adduct to an inorganic acid or base) to salvage the dualistic theory.[9] The four men named above, especially Wurtz and Hofmann, played a special role in what the historian Alan Rocke has termed the "quiet revolution" in organic chemistry.[10] Their distinctive styles of research, in which the ideas of Laurent and Gerhardt were developed, encouraged a relatively peaceful transition from those of Berzelius, Liebig, and Dumas to the ideas of Frankland, Williamson, Couper, and Kekulé.

Like Gerhardt, Wurtz was born in Strasbourg, received his early schooling there, and was fluent in both French and German. His father, a Protestant pastor, hoped that his son would study theology, but Adolphe chose chemistry. In 1842 he went to Liebig's laboratory for a semester, and received an M.D. degree at Strasbourg in 1843 for a dissertation on albumin and fibrin. While in Giessen, Wurtz met Hofmann, with whom he formed a close friendship that lasted to the end of their lives. In 1844, Wurtz went to Paris to seek his fortune, and in 1845 Hofmann went to London, where he headed the Royal College of Chemistry until 1865. They visited each other frequently. Wurtz's career in Paris was determined by the favor shown him by Dumas, who appointed him his *préparateur* in organic chemistry at the medical school, and his *répétiteur* at the *Ecole Centrale des Arts et*

Manufactures. In 1845, his father died, and the financial support he received from him stopped, obliging young Wurtz to supplement the modest income from his academic appointments. Like Gerhardt, who translated some of Liebig's writings into French, Wurtz translated Gerhardt's first textbook (*Précis de Chimie Organique*, 2 vols, 1844-1846) into German. Although he gained knowledge of the ideas of Laurent and Gerhardt from it, it appears that Wurtz recognized the danger of a close association with his ostracized fellow Alsatian. Perhaps, as was noted above, at that time Gerhardt was in Montpellier, and Dumas overlooked Wurtz's seeming indiscretion, for in 1852 Wurtz was named Dumas's successor as professor of organic chemistry and pharmacy at the medical school (he became the dean in 1866), but it was not until 1875 that he was made professor of chemistry at the Sorbonne.[11] In 1869 he published his *Histoire des Doctrines Chimiques depuis Lavoisier jusqu'à nos Jours*, in which a lengthy chapter is devoted to Laurent and Gerhardt.[12]

Apart from his discovery of ethylamine, and his friendly competition with Hofmann in the study of compounds of the ammonia type, Wurtz made numerous other organic-chemical contributions, including a study of glycol and its derivatives, a method for the synthesis of hydrocarbons by the treatment of a mixture of two alkyl iodides with sodium, and the discovery of the condensation of two molecules of an aldehyde to form an "aldol." He also returned to the biochemical interests of his student days, identified a new soluble enzyme (papain), and wrote a valuable biochemistry textbook.[13]

After 1853, Wurtz attracted to his laboratory Archibald Scott Couper and Maxwell Simpson from Britain, Aleksandr Butlerov from Russia, J. H. van't Hoff from Holland, and several Germans, notably Friedrich Beilstein and Albert Ladenburg. His French research students included Edouard

Grimaux and his fellow Alsatians Charles Friedel and Joseph Achille Le Bel. He was much admired by his research students for his encouragement of individual initiative, and by the medical students for the clarity of his lectures and his liberality as an administrator.[14] Wurtz's international outlook led him to recognize the importance of the work of Alexander Williamson (who had been in Paris during 1846-1849) and to establish a relationship with August Kekulé, who spent a year (1851-1852) in Paris.

During 1853-1859, Wurtz underwent a gradual conversion to the Laurent-Gerhardt definition of atom, molecule, and equivalent, adopted Gerhardt's atomic weights, and became the leading French proponent of chemical atomism. In this stance, he encountered the rivalry of Marcelin (or Marcellin) Berthelot, who became professor of chemistry at the pharmacy school in 1859, and at the Collège de France in 1865. As an organic chemist, Berthelot made numerous contributions through his work on the synthesis of hydrocarbons, and on sugars, glycerol derivatives, terpenes, and camphor. With Péan Saint-Gilles, he studied the kinetics of the esterification of alcohols, and then turned to research in thermochemistry. In the interpretation and presentation of his experimental results, Berthelot declined to accept the atomic theory, or the ideas of Laurent and Gerhardt, and clung to the use of equivalents until the 1890s. In 1877, Berthelot and Wurtz debated atomism before the Académie des Sciences. In this contest, Berthelot was clearly the winner, so far as the teaching of chemistry at French schools and colleges was concerned. Toward the end of the century, Berthelot studied Greek alchemical writings, and was active in French politics, with a brief spell (1895-1896) as Foreign Minister. After the defeat in the Franco-Prussian war, and the loss of Alsace, both Wurtz and Berthelot emphasized the past greatness of French chemistry, but Wurtz did so in an international spirit, whereas Berthelot was more chauvinistic in that regard.[15]

August Wilhelm Hofmann was born in Giessen. His father was a building contractor, and in 1839 had enlarged and rebuilt Liebig's laboratory. Young Wilhelm studied law and languages at the university, but shifted to chemistry after attending Liebig's lectures. He received his doctoral degree in 1841 for a dissertation on the volatile components of coal tar, especially aniline and quinoline. In the continuation of this work, Hofmann was helped by Laurent, who was at Liebig's laboratory for a brief visit during 1843. As was noted above, Hofmann's studies on derivatives of aniline obliged Laurent to change his classification of aniline from belonging to a phenyl type to one belonging to an ammonia type.

In 1844, Hofmann became Privat-Dozent in Bonn, but in he following year he moved to London to be professor at the newly established Royal School of Chemistry. In accepting this appointment, he obtained from the Bonn administration a leave of absence, which was renewed several times afterward. As it turned out, Hofmann did not return to a German university until 1865, and it was not Bonn, but Berlin. The attractions of London were many, not the least of which was a much higher salary, allowing Hofmann to marry Helene Moldenhauer (Liebig's niece) and to enjoy the company of a remarkable group of British chemists, as well as that of Kekulé, who worked in London during 1854-1855. The financial difficulties of the privately supported College led to its transfer, in 1853, to the Government School of Mines, and the appointment of Hofmann as Lyon Playfair's successor as Director.[16]

Like Wurtz, Hofmann had adopted the atomic weights of Gerhardt and Cannizzaro by 1860. During the 1850s he extended his studies on organic compounds of the ammonia type, discovered the ammonium compounds (e.g., tetraalkyl ammonium iodide), and introduced the method of "exhaustive methylation." Later, he found a method for converting acid amides to amines by treatment with bromine and alkali; a method that became known as a "Hofmann reaction." He also collaborated with Auguste Cahours in extending the concept to the ammonia type to the analogous phosphine (PH_3) derivatives, and in work on allyl alcohol. In addition to studies on isocyanates and alkaloids, Hofmann had an abiding interest in synthetic dyes; his enterprising 18-year-old student, William Henry Perkin, working in his laboratory at home, accidentally discovered "aniline purple" (also known as mauve or mauveine) and, much to Hofmann's displeasure, patented it and left Hofmann to set up a factory for its manufacture. Hofmann himself developed several new dyes of the triphenylmethane type; they were later studied intensively by Emil Fischer.[17]

All accounts of Hofmann's style attest to his enthusiasm and finish as a researcher and teacher. As Henry Armstrong put it:

> The finished style of his early work, which presumably was entirely his own, appears all the more remarkable when we bear in mind that—as all who were associated with him tell us—he was himself a most awkward manipulator; and that the success he afterwards achieved was due to his extraordinary power of inciting others to work, and to the remarkable judgment he displayed in selecting helpers gifted with powers of experimenting with which himself he was so indifferently endowed. And the manner in which he returned his thanks was equally striking and characteristic; indeed, if the complimentary references in his papers to those who at various times assisted him were collected, they alone would afford an interesting and matchless record of his diplomatic talents and courtly manners.[18]

Moreover, Hofmann's papers were singularly free of the offensive tone that frequently marked those of Liebig and his adversaries. Hofmann's devotion to Liebig is all the more remarkable, because no sooner had Hofmann settled in London than Liebig began to involve him in some of his various financial interests in England, notably a quarrel with John Bullock and John Gardner about the distribution of profits from the sale of a quinine preparation.[19] When Liebig learned in 1863 that Hofmann was planning to return to Germany, he urged Hofmann to remain in London because "England is so extremely poor in truly scientific men that a man such as you is a true blessing for the country; in no other can he hope to achieve more or to gain greater influence."[20] Soon after his arrival in Berlin, Hofmann promoted the founding of the *Deutsche Chemische Gesellschaft* as a counterpart of the Chemical Society, and of its *Berichte*, to serve as a more civil and less personal medium of chemical publication than Liebig's *Annalen* or Kolbe's *Journal für praktische Chemie*. Hofmann also encouraged the growth of the German synthetic dye industry, as he had tried to do in England.

Among Hofmann's chemical colleagues in London, Alexander Williamson stands out as second only to Kekulé, who acknowledged Williamson's influence on his thought. Williamson had grown up in a well-off, refined family, but his childhood had been marred by the malpractice of physicians who caused him to lose much of his vision and the use of his left arm. He overcame these lasting handicaps with admirable determination. During his youth, Williamson attended schools in London, Dijon, and Wiesbaden, and in 1841 enrolled in the medical course at Heidelberg. There, Leopold Gmelin excited his interest in chemistry. After gaining his parents' acceptance of his decision to become a chemist, and to set up a private laboratory in their home, Williamson spent two years (1844-1846) as a student in Liebig's laboratory. His eagerness to do original experimental work was chastened by Liebig's insistence that he do some routine analyses first, but before Williamson left Giessen, he had published several chemical papers on hypochlorous acid and Prussian blue, and was granted a Ph.D. without having to submit a thesis.

On the advice of John Stuart Mill to his father, the young Williamson was sent to Paris to study mathematics with Auguste Comte. He spent three years there, fitted out a private laboratory (where he probably began his experiments on etherification), and met numerous French chemists, including Laurent and Gerhardt. Williamson also attended Comte's lectures on philosophy, and later showed strong philosophical leanings in his chemical thought, but he did not accept Comte's extreme positivist doctrine. In 1849, Thomas Graham encouraged Williamson to apply for the open professorship of practical chemistry at University College London. The numerous letters in support of his application provided by Liebig, Dumas, Hofmann, and others were effective, and later that year Williamson returned to London; in 1855, he succeeded Graham as professor of general chemistry, and the two professorships were combined. Although Willliamson did not retire until 1887, after 1855 his principal activities were in teaching, giving public lectures, and committee work. He also patented several pharmaceutical preparations and invented an improved marine boiler.[21]

In 1850-1851, Williamson published his elegant papers on the constitution of alcohols and ethers, and introduced the water type into the classification of organic and inorganic compounds. He disproved the theories offered by Mitscherlich, Berzelius, and Liebig to explain the conversion of ethyl alcohol to ether by first examining the action of ethyliodide (C^2H^5I, previously prepared by Edward Frankland) on potassium ethylate (C^2H^5OK). He found "to my astonishment, the compound that formed was nothing else but common aether, $C^4H^{10}O$." [22] He then showed that, in the action of sulfuric acid, the process involves two C^2H^5 successive reactions: (1) the formation of sulfovinic acid, which he formulated as $^{C^2H^5}_H so^4$; (2) the reaction of this product (which he considered to be comparable to ethyl iodide) with another molecule of ethyl alcohol to form $^{C^2H^5}_{C^2H^5} O$. Moreover, he demonstrated the formation of "mixed ethers" from ethyl alcohol and amyl alcohol. In a paper on the constitution of salts, Williamson wrote:

> When we study a molecule by itself, we study it physically; chemistry considers the *change* effected by its reaction with another molecule, and has to describe the process by which that change is effected. A chemical decomposition should therefore be represented by the juxtaposition of formulae of the reacting substances, and by effecting in these formulae the change which takes place in the mixture. The adoption of this method will of course necessitate the adoption of types, from which, by the replacement of certain elements or molecules, we can deduce the constitution of more complex groups. I believe that throughout inorganic chemistry, and for the best-known organic compounds, one single type will be found sufficient; it is that of water, represented as containing 2 H atoms of hydrogen to 1 atom of oxygen, thus $^H_H O$. [23]

In these papers, Williamson enunciated a dynamic theory of chemical reaction, and dismissed the speculations about "contact action" or "catalytic force" by Mitscherlich and Berzelius.

A steadfast adherent of the atomic theory, in 1869 Williamson gave a famous lecture in which he discussed at length the arguments for and against the theory. At the end of the lecture, he stated:

In using the atomic language and atomic ideas, it seems to me of great importance that we should limit our words as much as possible to statements of fact, and put aside into the realm of imagination all that is not in evidence. Thus the question whether our elementary atoms are in their nature indivisible, or whether they are built up of smaller particles, is one upon which I, as a chemist, have no hold whatever, and I may say that in chemistry the question is not raised by any evidence whatever. . . . In conclusion I must say that the vast body of evidence of the most various kinds, and from the most various sources, all pointing to the one central idea of atoms, does seem to me a truly admirable result of human industry and thought. Our atomic theory is the consistent general expression of all the best known and best arranged facts of the science, and certainly is the very life of chemistry.[24]

One of Williamson's chemical colleagues, Benjamin Collins Brodie, Jr., also had an excellent mathematical education and learned chemistry in Giessen, but they went separate ways in their chemical thought. A respected member of the British chemical community, Brodie had been elected a Fellow of the Royal Society in 1849, and was professor of chemistry at Oxford (1855-1872). In 1864, he presented as an alternative to the atomic theory his "calculus of chemical operations," by means of which he interpreted the results of his study of peroxide compounds. There ensued a lively discussion, with criticism of both his chemistry (especially by Alfred Naquet) and his mathematics.[25] Other efforts to bring mathematics into chemistry included those of the noted algebraist Arthur Cayley, who applied the graph theory of mathematical trees to chemical isomers, and of James Joseph Sylvester, Cayley's collaborator in the development of invariant theory, who attempted to apply that theory to chemical problems.[26] More recently, Joshua Lederberg described the use of graph theory in the mapping of organic molecules.[27]

Among Williamson's chemical friends, his former pupil William Odling was more significant than Brodie. A scion of a medical family, Odling received his M.D. in 1851, after studies at Guy's Hospital. He taught chemistry there until 1855, and afterward at St. Bartholomew's Hospital. In 1867, he succeeded Faraday at the Royal Institution, and in 1872 he succeeded Brodie at Oxford, where he remained until his retirement in 1912. Odling was less inclined to laboratory work than to theorizing, but the clarity of his papers and lectures made him an effective defender of the theories of Laurent, Gerhardt, and Williamson. Apart from his translation of Laurent's *Méthode de Chimie*, Odling's chief personal contributions were papers published during the 1850s, when he introduced a notation to indicate what was later termed "valency" and suggested the addition of a "marshgas" (methane, CH^4) type to those proposed by Gerhardt.[28]

The central figure in the chemical revolution of the 1850s was, of course, August Kekulé, who was in London during 1854-1855, and of whom much more shortly. He frequently acknowledged the stimulus he derived from his association with Williamson and Odling. Before discussing Kekulé's achievements, the important contributions of two other British chemists—Edward Frankland and Archibald Scott Couper—must be taken into account.

Born an illegitimate child, a fact made public only after his death, Frankland emerged from a Lancastrian background of poverty and interrupted elementary schooling to prominence in Victorian society through his achievements in chemical research and education, and in the promotion of public health.[29] Frankland's chemical work and thought received stimulus from various sources—his association in London and Marburg with Hermann Kolbe, who had remained attached to Berzelius's dualistic doctrine, a subsequent stay in Marburg, during which Frankland appreciated more fully Bunsen's chemical skill and technical inventiveness, and the influence in Giessen of Liebig's associate Heinrich Will, who had adopted much of Gerhardt's theoretical approach.[30]

Frankland began his chemical education as a pharmacist's apprentice. In 1845 he went to London, where he worked in Lyon Playfair's laboratory in Putney, and where he met Kolbe, whom he taught English in exchange for lessons in German, and with whom he published an important paper on the conversion of alkyl cyanides to fatty acids. They continued this work in Bunsen's laboratory in Marburg, where Frankland stayed only three months before taking up a teaching post at Queenswood College in Hampshire. He found the post to be uninviting but, inspired by Bunsen's work on "cacodyl," he joined in the hunt for the elusive ethyl radical. A year later, he returned to Marburg (in the company of John Tyndall) and in 1849 received his Ph.D. for a dissertation entitled *Ueber die Isolierung des Aethyls*. Frankland then spent the winter of 1849-1850 in Giessen, and had planned to go on to Berlin, but was summoned by Playfair to return to London and accept an appointment that led to Frankland's designation as Playfair's successor as professor at the Putney College of Civil Engineering.

At Putney, Frankland discovered that sunlight promotes the reaction of ethyl iodide with zinc to produce what he called "zinc ethyl" and he introduced the term "organometallic" chemistry.[31] In his papers on this subject he did not use the modern atomic weights, which lead to the formula of zinc ethyl as $Zn(C_2H_5)_2$, not Frankland's $Zn(C_2H_5)$. He also combined the radical theory with the substitution type theory, and used the terms "saturation capacity" and "combining power" to denote the maximum number of substituent groups that could be accepted by an element. Such terms were adopted in 1857 by Kolbe in a paper of which Frankland felt that he should have been a co-author. After the appearance of Kekulé's famous 1858 paper [32] and the first installment of his *Lehrbuch der organischen Chemie* (1859), in which he presented his idea of "atomicity," Frankland claimed priority in the development of the concept of valency. The issue was never fully resolved, and although Kekulé has generally been given the chief role in that development, Frankland's contribution has received due recognition. [33]

Frankland remained at Putney for only one year. In 1851 he was appointed professor of chemistry at Owens College Manchester. He did not find a climate conducive to research there, and in 1857 he returned to London, where he was given a laboratory at St. Bartholomew's Hospital. In 1865 he succeeded Hofmann as director of the Royal College of Chemistry, where he remained until his retirement in 1885. At the Royal College, Frankland was joined by Baldwin Francis Duppa, a well-trained chemist who had worked with Hofmann and Perkin (Duppa died in 1873). Work was continued in organometallic chemistry, but then shifted to the use of synthetic methods in the study of derivatives of lactic acid and of oxaloacetic acid. Frankland's other research publications included a report of calorimetric data in support of the disproof, by Adolf Fick (his brother-in-law) and Johannes Wislicenus, of Liebig's view that muscular contraction is effected by the decomposition of muscle protein. [34] After 1865, Frankland was active in the field of public health, and provided improved methods for testing the chemical contamination of the water supply. He was aided in this effort by Henry Armstrong.

If Frankland's scientific career ended with public acclaim, that of Couper was cut short at its beginning. The son of a wealthy Scottish cotton manufacturer, Couper first studied classics and philosophy at Glasgow. He traveled on the

continent, and in 1855, while in Berlin, he was attracted to
organic chemistry by the lectures of Franz Sonnenschein and
Karl Rammelsberg. Couper then worked during 1856-1858 in
Wurtz's laboratory, and his first publication (1857)dealt with
his preparation of bromobenzene and dibromobenzene by
the bromination of benzene. Couper then turned to the reaction of salicylic acid and methyl salicylate with phosphorus
pentachloride. In 1858 he reported the formation of trichlo-
rophosphosalicylate, and the decomposition of this product
to the monochloro compound and to phosphosalicylic acid.
In one of his papers on this subject, Couper wrote constitu-
tional formulas in which the linkages between atoms were
denoted by means of dotted lines; for these formulas, he
used C = 12, O = 8. During 1858, Couper also composed a
paper entitled "Une nouvelle théorie chimique," based on
the affinity of an element in its combination with another
element in definite proportions, and the affinity of different
elements in their combination with one another. He proposed that carbon can combine with four substituents and
that its atoms can combine with one another to form chains.
The argument in that paper was strikingly similar to that in
Kekulé's paper, which appeared on 19 May. There were some
differences, however. Kekulé wrote of the "unequal basicities
of atoms" while Couper invoked "affinity of degree" or the
"limits of combination which the elements display." Whereas
Kekulé adopted some features of the type theory, Couper
rejected Gerhardt's theory altogether, because "[Gerhardt]
is led not to explain bodies according to their composition
and inherent properties, but to think it necessary to restrict
chemical science to the arrangement of bodies according to
their decomposition, and to deny the possibility of our com-
prehending their molecular constitution." [35] In early May,
Couper presented this paper to Wurtz for communication to
the Académie des Sciences, but Wurtz (who was not yet a

member) delayed asking Dumas to sponsor it, and the paper was not sent to the Académie until 14 June, after the appearance of Kekulé's famous paper on 19 May.[36] According to testimony later collected by Richard Anschütz, Couper became distraught and insulted Wurtz, whereupon Couper was dismissed from Wurtz's laboratory.[37] After a brief service as an assistant of Lyon Playfair, Couper was obliged to enter a mental hospital. He then returned to his parents' home in Scotland, and spent the rest of his life (he died in 1892) under the loving care of his mother.[38]

It should be added that between 1858 and 1885, numerous chemists (including Kekulé and Kolbe) had reported their failure to reproduce Couper's results on the chlorination of salicylic acid, but in the latter year Richard Anschütz (a former pupil and later biographer of Kekulé) confirmed Couper's claims by following closely Couper's directions. In the intervening years, Couper's work and theory were rarely mentioned and in 1890, at a meeting arranged by A. W. Hofmann to honor Kekulé, the latter reminisced about the early development of the valence theory, told of his debt to Laurent, Dumas, Gerhardt, and Williamson, belatedly acknowledged the role of Frankland, but did not mention Couper's name.[39] The confirmation of Couper's work on salicylic acid led Anschütz to enlist the help of Alexander Crum Brown (of whom more later) in seeking biographical data about Couper, and to correct the historical record.

In 1860, as the moving spirit at the famous Karlsruhe congress, Kekulé was the widely acknowledged leader of the young chemists who were fashioning the new structural organic chemistry. Only ten years earlier he had been a pupil of Liebig, Will, Strecker, and Fleitmann at Giessen, after having switched from architecture to chemistry. His academic future was uncertain, for Liebig questioned his ability as an experimenter and deplored his preference for theory. In his work with Liebig, Kekulé was assigned routine analyses and the analysis of gluten and wheat bran.[40] After receiving his doctoral degree in 1852 for work with Will on amyloxysulfates, Kekulé went to Paris, where he learned much from Gerhardt, and then spent 1852-1853 as the assistant of Adolf von Planta (a former Liebig student) who had a private laboratory near Chur, Switzerland. There, Kekulé worked on alkaloids and analyzed minerals and mineral waters, tasks not much to his liking. Liebig advised him to become a teacher at an agricultural school, but because he wanted to remain in chemical research, Kekulé accepted an assistantship with John Stenhouse (another Liebig pupil) at St. Bartholomew's Hospital in London. Liebig had recommended him for this post, but did not write Kekulé about the terms of the appointment; instead he had his teaching assistant, Wilhelm Mayer, do so. As Kekulé later recalled:

> I had little inclination to accept it because, if I may be permitted the expression, I considered him [Stenhouse] a *Schmierchemiker*. By chance Bunsen came to Chur to visit a brother-in-law whom I had come to know. I asked him about Stenhouse's offer, and he advised me strongly to accept it. I would learn a new language, but would not learn chemistry. I therefore went to London, where I did not profit much from my assistantship with Stenhouse.[41]

As it turned out, Kekulé's stay in London (1853-1855), apart from routine analyses for Stenhouse, led to his acquaintance with Williamson and Odling, whom he met frequently, and to occasional meetings with Frankland and other British chemists. In 1854 Kolbe launched an attack (of which more shortly) on the theories of Gerhardt and Williamson, and Kekulé aligned himself with their concepts. In that year, he published a paper on the action of phosphorus pentasulfide on acetic acid to produce thioacetic acid, and concluded that, as for oxygen, one atom of sulfur is "dibasic" and equivalent to two atoms of chlorine. He later claimed that what he called "atomicity" represented the first statement of the principle of valency.[42] In making that claim, Kekulé chose to omit mention of the earlier contributions of Frankland, and it was not until 1890 that he made the oft-quoted statement that "originally a pupil of Liebig, I became a pupil of Dumas, Gerhardt, and Williamson; I no longer belonged to any school."[43]

In 1854, Kekulé applied for a professorship at the new polytechnic school in Zürich but failed to get the support of Liebig, and the appointment went to Wöhler's assistant Georg Staedeler, whom Liebig had recommended. On his return to Germany in 1855, Kekulé became a *Privatdozent* at Heidelberg, where he set up a small private laboratory and a lecture room in which he taught modern organic chemistry to a group that included Adolf Baeyer, Emil Erlenmeyer, Lothar Meyer, and Henry Roscoe. Kekulé's famous 1858 paper and the first installment (1859) of his textbook emerged from this effort.[44] One of the first publications from his Heidelberg laboratory dealt with the preparation and elementary analysis of glycogen, just discovered by Claude Bernard. Kekulé's main experimental interest, however, was in the constitution of mercury fulminates, and in his student Baeyer's work on cacodyl compounds.[45]

In 1858, he became professor of chemistry at Ghent; Liebig recommended him as an excellent teacher, and characterized his research papers as "bearing the mark of conscientiousness, skill, and acuteness." During his stay in Ghent (1858-1867) Kekulé developed his benzene theory, proposed the holding of the Karlsruhe congress, and attracted to his laboratory, in addition to Baeyer, Albert Ladenburg, Wilhelm Körner, Hermann Wichelhaus (who introduced the term *Valenz* in 1868), and the Englishmen James Dewar and George Carey Foster.[46] At Ghent, Kekulé worked on a variety of problems, among them the chlorination of salicylic acid, the bromination of succinic acid, the isomerism of maleic and fumaric acids, and the constitution of aromatic and diazo compounds.

In 1867, Kekulé finally received a professorship at a German university. The post in Bonn became vacant in 1863, Hofmann was expected to return there, and a fine new institute was promised, but he went to Berlin, where an even finer building was provided. Kolbe was offered the post, but declined, and it was given to Kekulé. He lived out his years in Bonn, where his group did extensive work on aromatic compounds, with special reference to the isomerism of benzene derivatives. The most important experimental work on this question, however, was done by his former student, Körner at Milan; he synthesized over a hundred benzene derivatives, and provided the most convincing evidence for the hexagonal planar structure of benzene.[47]

Kekulé's initial graphical representation of benzene and its derivatives in 1866 was in the form of sausage-like structures, although he also offered hesitantly a hexagonal (cyclohexatriene) structure with alternating single and double bonds[48] (Fig. 6). Laurent's prior use of a hexagon to denote benzene was not mentioned, and Kekulé scornfully dismissed as "Confusionsformeln" the symbolism proposed in 1861 by Joseph Loschmidt, an Austrian schoolteacher better known for his contributions to physical theory.[49] After Ladenburg pointed out in 1869 that the cyclohexatriene structure was inconsistent with the fact that only one ortho-disubstituted benzene was obtained, Kekulé suggested that the vibration and collision of adjacent atoms in the molecule might lead to an oscillation between the two possible cyclohexatriene structures, thus anticipating the formulation of

Pour faire comprendre ce remplacement réciproque de
deux restes, je supposerai que, dans l'ammoniaque et le
chlorure de benzoïle, les atomes sont disposés suivant les
figures hexagonales :

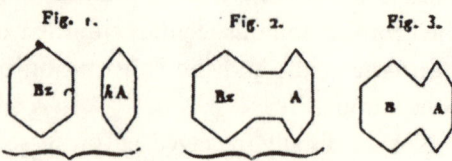

Laurent's depiction of the reaction of
bromobenzene with ammonia (1854).

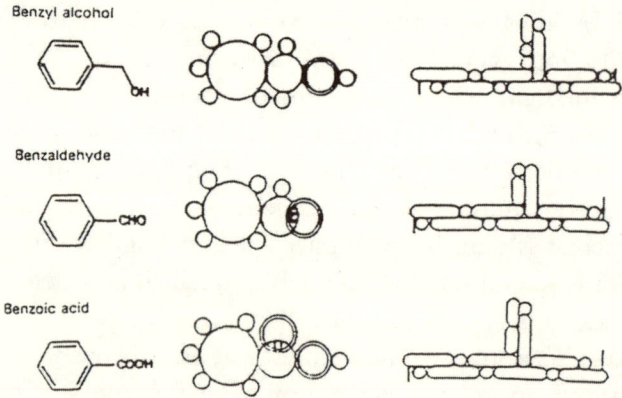

Loschmidt's (1861) and Kekulé's (1866) notation for benzene derivations.

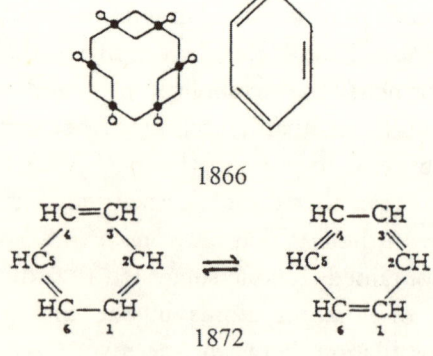

Kekulé's notations for benzene.

Figure 6. Graphical representation of benzene and benzene derivatives.

the concept of "resonance" some 50 years later. Alternative structures were proposed by Adolf Claus, Albert Ladenburg, James Dewar, Henry Armstrong, Adolf Baeyer, and Johannes Thiele (Fig. 7).

At Bonn, Kekulé's research group also worked on triphenylmethane, terpenes, the condensation of aldehydes, and

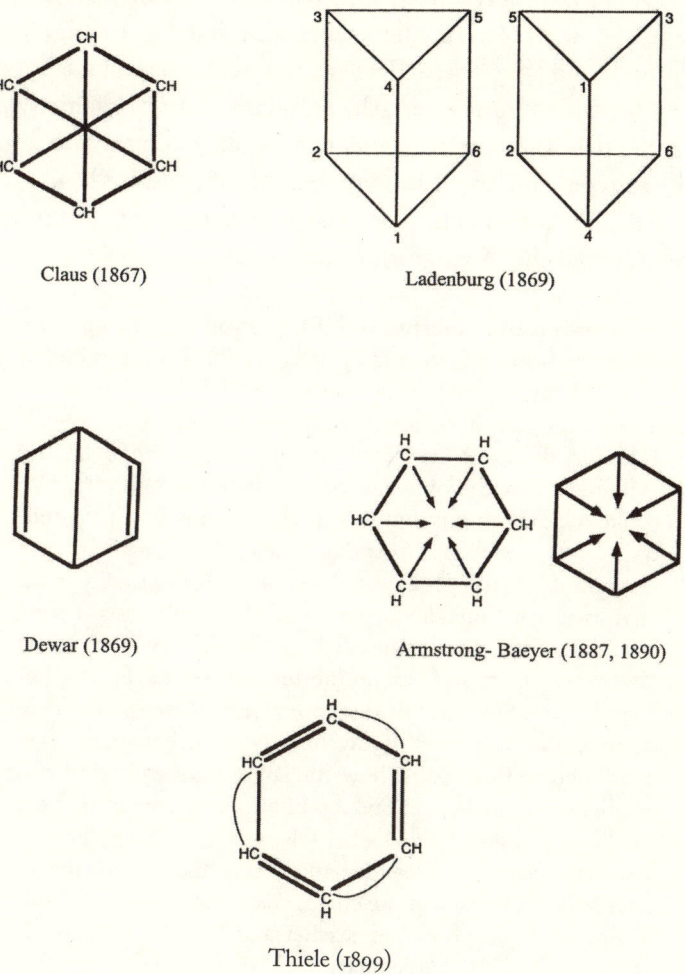

Claus (1867)   Ladenburg (1869)

Dewar (1869)   Armstrong- Baeyer (1887, 1890)

Thiele (1899)

Figure 7. Alternative structures for benzene.

the constitution of pyridine. In his paper on the condensation of aldehydes, Kekulé suggested (as had Baeyer before him) that such reactions are involved in the chemical activity of plants.[50]

From the testimony of his former assistants and students at Bonn, it appears that after about 1870, Kekulé's ailments led to increasingly infrequent visits to the laboratory.[51] In 1874, he was offered the chair occupied by Liebig, but declined, and proposed Baeyer instead. In Baeyer's laboratory (of which more later), the emphasis was on organic synthesis, based on the concepts of valence and structure, and the stereochemistry of Pasteur, van't Hoff, and Le Bel.

At the 1890 meeting in his honor, mentioned earlier, Kekulé told the distinguished audience:

> Perhaps it is of interest to you if I give you highly indiscreet accounts from my mental [geistigen] life how I came to some of my ideas. During my stay in London I lived for a while in Clapham Road, near the Common. However, I frequently spent evenings with my friend Hugo Müller in Islington at the opposite end of the enormous city. We spoke of many things, but mostly about our beloved chemistry. On a beautiful summer day I traveled once again on the last omnibus through deserted streets of this normally lively city; as usual, "outside" on the roof of the omnibus. I sank into dreaming. The atoms flickered before my eyes. I had always seen them in constant motion, these small bodies, but I could never succeed in discerning [erlauschen] the mode of their motion. Now I saw how two smaller ones often combined to form pairs; how the larger ones embraced two smaller ones, how larger ones bound three or even four of the smaller ones, and how it all revolved in a swirling dance. I saw how the larger ones formed a string and pulled smaller ones only at the ends of the chain. I saw what the past master Kopp, my highly honored teacher and friend, described so attractively in his "Molekularwelt," but I saw it long before him. The call of the conductor "Clapham Road" woke me

from my dreams, but I spent the rest of the night at least to put to paper sketches of these dream visions. So originated the structural theory.

It was likewise with the benzene theory. During my stay in Ghent in Belgium I occupied an elegant bachelor flat in the main street. My work room was however in a narrow side street and received no light during the day. For a chemist who spends the daylight hours in the laboratory that is not a disadvantage. I sat there and wrote at my textbook; but it did not go well; my mind was on other things. I turned the chair toward the fireplace and fell into a half-sleep. Again the atoms flickered before my eyes. Smaller groups remained this time modestly in the background. My mental eye, sharpened by repeated visions of this sort, now distinguished larger shapes of various configuration. Long chains, often joined more compactly; all in motion, winding and turning like snakes. And look, what was that? One of the snakes seized its own tail and the vision twirled scornfully before my eyes. I woke as if struck by lightning; this time also I spent the rest of the night in order to work out the consequences of the hypothesis.[52]

These recollections have recently been the subject of lively argument among chemists, historians, and psychologists.[53] In a paper dealing with the benzene story, Alan Rocke stated that:

Although its authenticity has occasionally been questioned, most historians are content to accept the story at face value. Indeed the amount of detail included in both this and the 'structure story' anecdote suggests that they probably happened pretty much as Kekulé described them, nor do we have any persuasive grounds to accuse him of deliberate historical falsification. Moreover, it would appear that Kekulé related at least the benzene story many times to friends and family before its publication near the end of his life.[54]

It has also been concluded that "Psychologists have cited the dream accounts in support of their preconceived theories, rather than deducing any important novel theories for it. Occasionally they further complicated it with factual inaccuracies."[55]

For the purpose of this essay, the veracity of these accounts, no matter their detail or frequency of their prior recital, is less important than the telling itself. In London and Ghent, we are told, Kekulé was "enterprising, poor, and ambitious."[56] Like many chemists before and after him, he tried to assert his priority on matters he considered to be important. We have seen from the previous discussion of the early development of valency theory that although he was ready to acknowledge the role of Gerhardt and Williamson, he was less than generous in his treatment of Frankland, Odling, and Couper. What may be understood and forgiven for a young chemist uncertain about his future is difficult to excuse for a widely acclaimed scientist near the close of a distinguished career.

At the Karlsruhe congress, organized by his former student, Carl Weltzien, Kekulé indicated his concern about the notation used in writing formulas.[57] In the first installment (1859) of his textbook, he had introduced a sausage-like symbolism to denote the arrangement of the atoms in acetic acid, and did not adopt until 1867 the linear designation used by Couper and developed by Couper's friend, Alexander Crum Brown.[58] (Fig. 8). As was noted above, Kekulé rejected the symbolism proposed in 1861 by Loschmidt, who used circles and ellipses of various size to denote the individual atoms, but displayed great acumen in depicting the arrangement of atoms in many organic molecules.[59]

Much has also been written about the relative importance of the contributions of Kekulé and Aleksandr Butlerov to the formulation of the concept of "chemical structure."[60] This term was widely adopted after it was introduced by Butlerov in 1861:

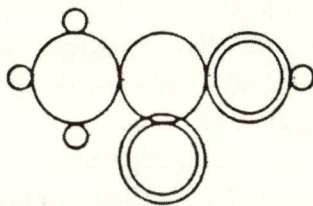

Couper's notation for butyl alcohol (1858).

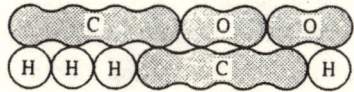

Loschmidt's notation for acetic acid (1861).

Kekulé's notation for acetic acid (1861).

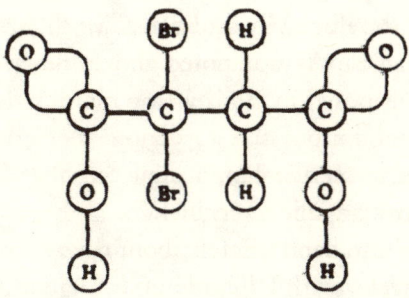

Crum Brown's notation for dibromosuccinic acid (1861).

Figure 8. Representations of organic molecules (1858–1861).

> Starting from the assumption that each chemical atom pos-
> sesses only a definite and limited amount of chemical force
> (affinity) with which it takes part in forming a compound,
> I might call the chemical arrangement, or the type and
> manner of the mutual binding of the atoms in a compound
> substance by the name of "chemical structure."[61]

The phrasing of this definition suggests the influence of
Couper, whom Butlerov met during a stay in Wurtz's laboratory in 1857. Although Butlerov repeatedly acknowledged
the importance of Kekulé's contributions, he criticized his
continued adherence to Gerhardt's type theory.

During the 1860s, however, German chemists considered
the term "chemical structure" to be solely attributable to
Kekulé. What I find most interesting about this attribution
is that Kekulé accepted it in silence, and did not mention
Butlerov in his published writings.

Perhaps the most incisive appraisal of Kekulé's style was
that offered in 1905 by his devoted pupil, Adolf Baeyer:
"Kekulé had no interest in substances themselves, but was
only concerned whether they conformed to his ideas. When
that was the case, it was fine; if not, they were rejected."[62]

I turn next to the role and style of Hermann Kolbe,
Kekulé's principal (and implacable) scientific adversary, an
outstanding experimenter and a leading figure (along with
Strecker, Frankland, Hofmann, and Perkin) in the emergence, after 1850, of synthesis as the main practical business
of organic chemistry.[63]

As was mentioned earlier, during 1845-1847 Kolbe worked
in Lyon Playfair's laboratory, where he began his collaboration with Frankland. Before coming to London, Kolbe
studied chemistry with Wöhler in Göttingen (1838-1842) and
received his Ph.D. at Marburg in 1843 as an assistant of
Bunsen. At that time, Berzelius was still alive and stubbornly
defending his dualistic radical theory against the attacks of

the French chemists, Liebig and he had parted company, and Wöhler and Bunsen were his principal protagonists. Kolbe thus inherited a veneration of Berzelius and, after about 1850, when both Wöhler and Bunsen had largely withdrawn from debates about organic chemical theory, it was Kolbe who became the chief champion of the dualistic doctrine.

During 1847-1851, Kolbe was engaged in editorial work for the Vieweg publishing firm in Braunschweig, but in the latter year he returned to academic life as Bunsen's successor at Marburg, after A. W. Hofmann had declined the invitation to accept that post. Despite initial difficulties, Kolbe's reputation as a lecturer and in the supervision of practical work grew rapidly, and he attracted outstanding students from Germany (among them Jacob Volhard, Carl Graebe, Edmund Drechsel) and abroad. The latter included Alexander Crum Brown, Aleksandr and Konstantin Zaitsev, and the American Charles W. Eliot. Kolbe's laboratory facilities at Marburg were greatly improved in 1863, but two years later he was appointed professor of chemistry at Leipzig, where a splendid new institute and private residence were constructed for him. There, among the students were such future notables as Ernst von Meyer, Ernst Beckmann, Theodor Curtius, Henry Armstrong, and Vladimir Markovnikov.[64] In 1870, Kolbe took over the editorship of the *Journal der praktischen Chemie*, a post he held, along with his professorship, until his death in 1884.

During the 1840s, Kolbe's experimental work was largely devoted to the study of organic acids. At that time, Berzelius had modified his electrochemical doctrine to include Gerhardt's "copula" theory, and considered organic acids to be constituted of a hydrocarbon radical paired (copulated) with hydrated oxalic acid. With great skill, Kolbe proceeded to effect several processes whose results he offered in support of

Berzelius's ideas. In a series of steps starting from the chlorination of carbon disulfide, Kolbe effected the synthesis of
acetic acid. Electrolysis of sodium valerate caused the release
of what he considered to be the "valyl" radical. With Frankland, he demonstrated the conversion, by alkali, of methyl
cyanide to acetic acid. Although the interpretation of this
conversion as involving the formation of oxalic acid from the
cyano group proved to be incorrect, the reaction itself was
an important addition to the methods of synthetic organic
chemistry. Another valuable by-product of Kolbe's work was
his prediction that, in addition to the existence of such compounds as butyl alcohol, there should an isomeric isobutyl
alcohol (a "secondary" alcohol) and a "tertiary" *t*-butyl alcohol.
The formulas Kolbe wrote for them were of course replaced
by those introduced by Crum Brown, but Kolbe insisted on
using his modification of the formulas of Berzelius. For example, he was prepared to admit that an acetyl radical really
exists, but he formulated it as $(C_2H_3)C_2$ [C = 6], and made
the methyl group the point of substitution by chlorine, and
the other carbon as the point of chemical reaction. In Kolbe's
classification, based on Frankland's revision of the type theory,
alcohols, aldehydes, and related acids belonged to a carbonic
acid (carbonyl) type.[65]

Around 1860, Kolbe disputed Wurtz's work on the constitution of lactic acid and other compounds, such as glycol and
glycolic acid, which have more than one chemically reactive
group. Kolbe also worked on salicylic acid and on the constitution of aromatic compounds, but rejected all hypotheses involving valence bonds and benzene rings, and did not
accept the new atomic weights until 1868. He proposed that
benzene is a derivative of methane in which three of the
hydrogens have been replaced by CH (methine) units. However, this proposal was inconsistent with what was known
from the work of Körner and others about the isomerism of

benzene derivatives; it has been reported that many of his research students did not accept it.[66]

After his rejection during the 1850s of the theories of Gerhardt and Williamson, Kolbe turned upon Kekulé, who had criticized Kolbe's formulas in his textbook. With the increasing acceptance by German chemists of Kekulé's *Strukturlehre*, and rejection of his own theories, Kolbe's growing bitterness became evident in such articles as the one in which he denounced van't Hoff and Felix Herrmann (van't Hoff's German translator) as "two virtually unknown chemists, one at a veterinary school and the other at an agricultural institute" and another article deriding Kekulé's rectoral address on the aims of chemistry.[67] In 1881, a long article, in several installments, on Kolbe's participation in the development of theoretical chemistry appeared in his journal.[68] In that paper he wrote:

> The sober prudent scientist will tell [Kekulé] that the object for which he and the majority of modern chemists strive is a chimera, that we will never succeed in gaining a conception of the arrangement of the atoms in the molecule, and that chemists should set for themselves a more modest goal: the investigation of chemical constitution in the sense of Berzelius. The goal for which Kekulé strives, and which he considers accessible, is actually more inaccessible for us than the moon, for we can see the moon and determine its form, but atoms we cannot see, and their form is perceivable with none of our senses.[69]

Kekulé prepared a rejoinder and submitted it to Volhard, editor of the *Annalen der Chemie*, but accepted Volhard's advice to withdraw it from publication.[70]

By 1882, Kolbe had lost the support of former friends, including A. W. Hofmann, Edward Frankland, and Hermann Kopp. His vituperation regarding Kekulé, his xenophobia for French chemists, and his anti-Semitism toward Jews in the German chemical society bordered on paranoia.[71] Seen at this distance, Kolbe emerges as a rather typical middle-class prejudiced German of the Bismarck era, and as an outstanding experimenter whose scientific vision was tragically clouded by his attachment to outworn ideas of the past.

To conclude this section, it must be added that the adoption of Cannizzaro's atomic weights and the theory of valency spurred the formulation of Dmitri Mendeleev's first periodic table of the elements in 1869. He was preceded by many chemists, some of whom based their proposals on Pythagorean numerology, most notably J.A.R. Newlands, who arranged the elements in the order of their equivalents, introduced the term "atomic number," and proposed what he called the "law of octaves." Lothar Meyer's table, published in 1870, was almost identical to Mendeleev's. The table predicted the existence of "missing" elements and described their expected properties; the subsequent discovery of such elements (gallium, scandium, germanium) led to its general acceptance. Apart from later corrections, the most serious deficiency of the table turned out to be its failure to predict the existence of the atmospheric gases—argon, helium, neon, krypton, and xenon—discovered during the 1890s by Lord Rayleigh and William Ramsay.[72]

Chapter Seven

STEREOCHEMISTRY AND ORGANIC SYNTHESIS

ॐ

IN WHAT HAS GONE before, I have sketched the development of three fundamental ideas—composition, constitution, and structure—in the thought of successive generations of chemists during the eighteenth and nineteenth centuries. The first depended on the identification of new elements (Lavoisier's "simple substances" or Davy's "undecomposable bodies"), and the analytical determination of their proportions in chemical compounds. The second emerged from the accumulation of knowledge about the transformations ("metamorphoses") of organic compounds in their reactions with other substances (for example, substitution reactions). The third, usually expressed in terms such as "the arrangement of atoms in a molecule," was invoked by Wollaston in 1808:

> I am further inclined to think, that when our views are sufficiently extended, to enable us to reason with precision concerning the proportions of elementary atoms, we shall find the arithmetical relation alone will not be sufficient to explain their mutual action, and that we shall be obliged to acquire a geometrical conception of their relative arrangement in all three dimensions of solid extension.[1]

A similar view was expressed in 1823 by Chevreul.[2] Later, the arrangement of atoms was offered as a possible explanation of the phenomenon of isomerism, but was considered by some chemists (for example, Gerhardt and Kolbe) to be inaccessible to scientific inquiry. Although, as mentioned earlier, Ampère, Gaudin, Baudrimont, and Laurent found in crystallography the stimulus to speculate about the actual arrangement of atoms, and to deny the impossibility of solving that problem, their ideas were dismissed by Dumas, Liebig, and Berzelius, who strongly defended their own speculations. At mid-century the attitude of most organic chemists to the study of the physical properties (other than the melting or boiling points) was that expressed in 1861 by Butlerov:

> Together with Gerhardt we deny for the present the possibility of accounting for the position of the atoms in the interior of a molecule; it seems quite obvious that chemistry, which only deals with bodies in a state of transformation, is powerless to judge this mechanical structure, as long as physical investigations are not brought to bear on this question. . . . I am sure that no one would say that this will remain inaccessible to us even in the future. . . . but we cannot deny, putting the concept of *physical atoms* entirely to the side, that the chemical properties of a body are determined in particular by the chemical bonding [*Zusammenhang*] of the elements which form it.[3]

Moreover, in 1865, Kekulé stated that:

> I consider it important to distinguish the chemical atom and molecule from physical particles and molecules and consequently I believe that we must determine the relative atomic weights by chemical reasoning based on the study of composition and metamorphoses. Perhaps the chemical units will be found to be identical with particles of matter which behave as units in one or another physical phenomenon; nevertheless, nothing demonstrates the necessity of such an identity, and one cannot accept it *a priori*.[4]

It is perhaps fair to say that, with the notable exception of Hermann Kopp, at mid-century German organic chemists (including Kekulé and Kolbe) did not consider the study of the physical properties of organic compounds to be worthy of serious attention. In particular, the phenomena of crystallinity and optical activity (the capacity of some organic compounds to turn the plane of polarized light) were not deemed useful adjuncts to the study of the constitution of such substances. Bunsen and Kirchhoff used the spectrochemical method they introduced mainly for the discovery of new elements, and absorption spectroscopy was initially applied in England by George Gabriel Stokes and a group of investigators who were known jocularly as "chromatologists."[5]

In what follows, I discuss the background of the memorable discovery by Louis Pasteur of the correlation of crystalline structure and optical activity, its impact on the work of van't Hoff and Le Bel, and the subsequent development of knowledge about the arrangement of the atoms of a molecule in space or, as it came to be called, stereochemistry.

In 1811 François Arago invented the polarimeter, which was greatly improved in succeeding years.[6] It was used during 1812-1838 by his colleague, Jean Baptiste Biot, in an extensive study of the optical activity of various crystals, of the oils of turpentine, lemon, and laurel, and of solutions of camphor (in alcohol) and cane or beet sugar (in water). Biot also demonstrated the dependence of the magnitude of the optical activity on the color of the light (optical dispersion).[7]

Biot, professor of physics at the Collège de France, and of astronomy at the university, was a central figure in the background of Pasteur's discovery. During the 1830s, Biot examined two known forms of tartaric acid, one that had been isolated by Scheele, and the other later found by Charles Kestner, studied by Johann Friedrich John, and named "racemic acid" (also termed "paratartaric acid") by Gay-Lussac. Biot found that the first was dextrorotatory, while the other was optically inactive. During the early 1840s, these observations were extended to the examination of the salts of the

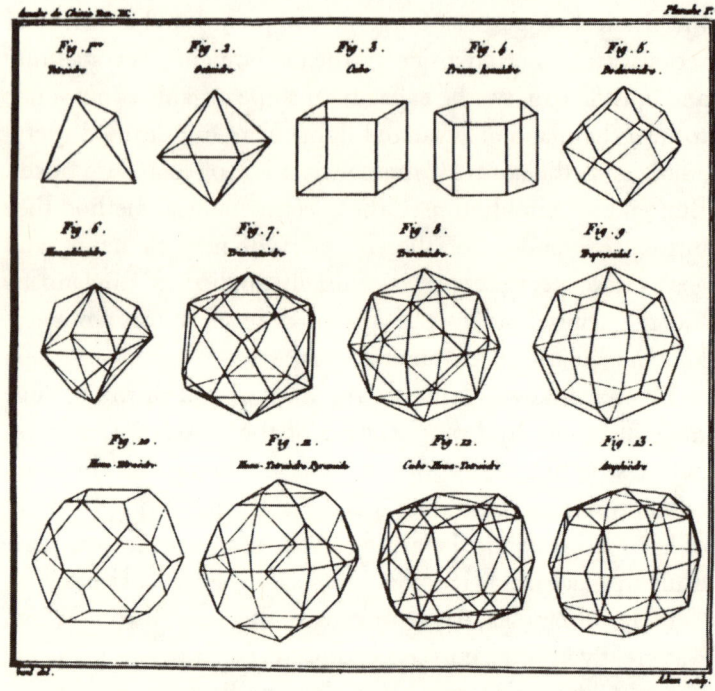

Figure 9. Models of molecular polyhedra (Ampère 1814).

two forms by Frédéric Hervé de la Provostaye, Remigius Fresenius, and Eilhard Mitscherlich; among these salts was sodium ammonium tartrate, the key substance in Pasteur's discovery.

In addition to Biot, two other people—Gabriel Delafosse and Auguste Laurent—played a large role in Pasteur's education. Delafosse, who introduced him to crystallography at the Ecole Normale Supérieure, was a pupil of Haüy, and at that time was seeking to refine and extend his teacher's idea of "molécules intégrantes" in the light of the speculations of Ampère[8] (Fig. 9). In particular, Delafosse called attention to the importance of an asymmetric feature noticed by Haüy, but not included by him in his geometrical system of classification. Haüy termed such crystals as "plagihedral"; later the term was changed to "hemihedral." In 1820, John Herschel discovered two hemihedral forms of quartz crystal, one of which he described as left-handed and the other as right-handed, and reported that they rotated the plane of polarized light in opposite directions[9] (Fig. 10).

In 1843, Delafosse suggested revisions of Haüy's assignment to the molécules intégrantes of some minerals of a cubic structure, and proposed instead a cubic assembly of tetrahedrons; he used the term "molécule" to denote the chemical entity, and "particule" for Haüy's term. As Auguste Bravais, one of the founders of the mathematical theory of crystallography, wrote soon afterward, "M. Delafosse had attributed hemihedry to differences in the structure of the molecule, neverthe-

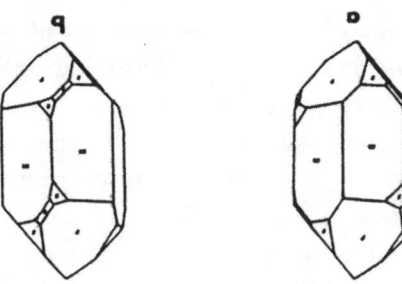

10a. Quartz crystal: a, levorotatory; b, dextrorotatory (Herschel, 1822).

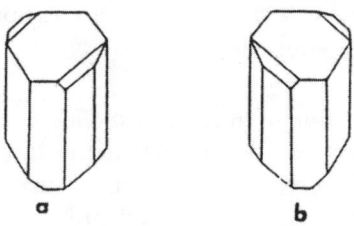

10b. Sodium ammonium tartrate: a, dexrorotatory;
b, levorotatory (Pasteur, 1850).

Figure 10. Hemihedral pairs of crystals

less without making known the general rules suited to enable us to pass from the knowledge of the [crystalline] faces suppressed by hemihedry to the determination of the figure of the molecule, or *vice versa.*"[10] Bravais' mathematical approach was further developed by Evgraf Federov, but its validity in relation to the actual structure of molecules was not established until after the introduction of the X-ray diffraction technique by von Laue and his associates.

In 1846, Pasteur began research for his doctorate in the laboratory of Antoine Jérôme Balard at the Ecole Normale, where Laurent worked during 1846-1847 after his return to Paris from Bordeaux. Laurent impressed Pasteur greatly; in March 1847 he wrote a friend: "M. Laurent is destined to occupy the premier position among chemists in a few years."[11] Laurent suggested a line of research related to his "nuclear" theory of organic-chemical constitution, and to his search for a connection between dimorphism and optical activity; in 1844, he had reported that strychnine and chlorostrychnine are isomorphous and have the same optical activity. Pasteur's work began with the study of the chemistry and crystal forms of arsenious acid, antimonious acid, and their salts, and shortly afterward he turned to the isomorphism of the tartrates, with special attention to their water of crystallization. In March 1848, he published a short paper on dimorphism, which concluded with a reference to an 1845 paper by Laurent on isomorphism, and stated:

> Finally, I add a proof in favor of the ideas which are [the] basis of this research, by an announcement of a remarkable fact which will soon be published in detail: namely, the first results of work that I undertook with MM. Courcière and Feuvrier, students at the Ecole Normale. We have found that the following eight tartrates, neutral tartrates of potassium, sodium, and ammonium, double tartrates of potassium and ammonium, of potassium and sodium, of sodium

and ammonium, and finally the bitartrates of potassium and ammonium are isomorphous, and can crystallize in all proportions. Nevertheless, these tartrates belong to different systems, the oblique rectangular prism, and the right rectangular prism.[12]

When, later in 1848, Pasteur found that all the tartrates were hemihedral, he examined the optically inactive sodium ammonium paratartrate, and he reported the discovery that it also was hemihedral (Fig. 10), but:

> I saw that the tetrahedral facets, corresponding to those of the isomorphous tartrates, were located relative to the principal faces of the crystal, sometimes to the right, sometimes to the left, in the different crystals which I had obtained. . . . I carefully separated the hemihedral right-handed crystals from the hemihedral left-handed crystals; I observed separately their solutions in M. Biot's polarimeter, and I saw, with surprise and delight [bonheur] that the right-handed hemihedral crystals turned the plane of polarization to the right, the left-handed hemihedral crystals to the left.[13]

From the longer papers that appeared in 1849 and 1850, it is evident that Pasteur had acquired considerable skill as a crystallographer, and that he was ready to draw far-reaching conclusions from his results. In continuing his program of research, he examined natural asparagine, aspartic acid, and malic acid, and found them to be optically active. But he was unable to establish the kind of correlation with hemihedry found with sodium ammonium tartrate. During the early 1850s, he also became interested in the chemistry of fermentation processes, with the finding that natural amyl alcohol is optically active, but he encountered difficulties in preparing good crystals for microscopic examination.

The circumstances of Pasteur's great discovery, so briefly reported in his 1848 paper cited above, were recounted by him more extensively in the first of two lectures in 1860 before the Chemical Society of Paris.[14] The text of these lectures has been used by numerous biographers and historians to describe Pasteur's research on the correlation of crystal structure and optical activity.[15] Further insight into the style of Pasteur's research on this subject has been gleaned from critical study of his research notebooks.[16]

Clearly, Mitscherlich and de la Provostaye either had missed seeing the hemihedral facets, or had disregarded them. It is possible that Pasteur's eyesight was keener than theirs, but it seems more likely that, thanks to Delafosse and Laurent, he was looking for hemihedry. Although Pasteur failed to confirm his generalization about the connection of hemihedry and optical activity, his work on amyl alcohol, asparagine, aspartic acid, and malic acid led him to the conviction that the "molecular dissymmetry" (as he called it in his 1860 lectures) of such optically active natural substances was correlative with their biological origin. I will return to this theory shortly. It should also be noted that sodium ammonium tartrate is exceptional among the paratartrate salts in the physical separation of the two optical isomers, and that this separation does not occur at temperatures above 27°C.[17]

During the course of his studies and research at the École Normale, Pasteur had weekly meetings with Dumas, his nominal supervisor, and probably came to understand that a continued association with Laurent might jeopardize his academic career. In February 1852, Pasteur wrote to Dumas thanking him for his counsel, and stated: "I had worked under the guidance of the good M. Laurent whom death is, perhaps, soon to remove him from science. I was at an age when the mind is fashioned on the model which is presented to it. I was enveloped by hypotheses without basis, by a redaction which completely lacked precision, and I spoiled the exposition of new and interesting facts. I was quickly enlightened by your counsel."[18] Laurent's name does not appear in any of Pasteur's published research papers after 1852.

A brilliant student at the school in Arbois, his birthplace, Pasteur spent two more years in additional preparation before being admitted in 1843 to the École Normale Supérieure, where he received his doctoral degree four years later. 1848 was a traumatic year for him. Apart from the tumult in the streets of Paris, Pasteur was obliged to interrupt his research in May because of his mother's fatal illness. To help his father in his months of grief, Pasteur accepted a teaching post at a lycée in nearby Dijon, but by the end of the year he was appointed acting professor of chemistry at Strasbourg; he was made a full professor in 1853. In 1855, Pasteur moved to Lille, to be dean of the science faculty, and it was there that he began his famous studies on fermentation, which marked a new stage in the development of microbiology. Finally, in 1857, Pasteur returned to Paris and to the École Normale, where he was director of scientific studies. In his administrative functions at Lille and Paris, Pasteur showed himself to be an efficient disciplinarian. Ten years later, he was named professor of chemistry at the Sorbonne and director of a laboratory of physiological chemistry. With each step on the academic ladder, and with his successive contributions to practical microbiology—silkworm disease, chicken cholera, anthrax, rabies—Pasteur's public fame grew, as did the financial support he received from Napoleon III and the post-1870 government, culminating in the establishment, in 1888, of the Pasteur Institute, of which he was director until his death in 1895.

During the preceding 35 years, Pasteur proved himself to be a formidable scientific debater, with Berthelot in 1860 about the nature of ferments and in 1877 about Claude Bernard's unpublished experiments on fermentation, with Félix Archimède Pouchet on the long-standing claims for spontaneous generation, with Liebig about the question of whether fermentation is a chemical or a microbial process, and with many critics of Pasteur's use of vaccines in medical and veterinary practice.[19]

I return to Pasteur's 1860 lectures on molecular dissymmetry. In the first lecture he stated:

> We know in effect, on the one hand, that the molecular structures of the two tartaric acids are dissymmetric, and on the other, that they are rigorously the same, with the only difference in exhibiting the dissymmetries in opposite senses. Are the atoms of the right-handed grouped on the spirals of dextrorotary helix, or placed at the summits of an irregular tetrahedron, or laid out according to some particular dissymmetric assembly or other? We cannot answer these questions. But what cannot be doubted is that there is a grouping of the atoms in a dissymmetric order with a nonsuperposable mirror image. What is no less certain is that the atoms of the left-handed acid are arranged precisely in a dissymmetric inverse grouping. Lastly, we know that paratartaric acid results from the juxtaposition of these two inversely dissymmetric groupings.[20]

In the second lecture he stated:

> All artificial products of the laboratory and all mineral species are superposable on their images. On the other hand, most natural organic products (I might even say all, if I were to name only those which play an essential role in the phenomena of plant and animal life), all the essential products of life, are dissymmetric, and possess such dissymmetry that they are not superposable on their images.[21]

In 1883, he returned to this theme in another lecture before the chemical society, and reiterated his oft-stated view that "there does not exist, to my knowledge, a single product of chemical synthesis, formed under the influence of causes which one may consider to belong to plant life, which is not dissymmetric, which does not have, in other words, the general form of a helix, a winding staircase, an irregular tetrahedron, a hand, an eye." In that lecture, Pasteur disputed the claim of Henry Perkin and Francis Duppa and of Emile Jungfleisch to have converted optically inactive synthetic succinic acid into optically active tartaric acid. Pasteur also mentioned that Joseph Achille Le Bel

> discovering, or rather divining, by means of ingenious theoretical ideas, the existence of various paratartrates, having in their formulas what he calls asymmetric carbon, for example the propylglycol of M. Wurtz, has also separated his paratartrates into right-hand and left-hand substances. . . . When the chemist combines elements in his laboratory he only applies non-dissymmetric forces. That is why all the syntheses he achieves never have dissymmetry.[22]

There is no mention in that article, nor in any other paper by Pasteur, of van't Hoff's name, and Le Bel's "ingenious theoretical ideas" were probably deemed to have some resemblance to those of Laurent. Before considering the memorable contributions of van't Hoff and Le Bel, I call attention to Pasteur's statement that: "You know that the most complex molecules in plant chemistry are the albumins. Moreover, you know that these immediate principles have never been obtained in a crystalline state."[23] Apparently, Pasteur did not know, or chose to ignore, the crystallization of plant seed proteins by several German investigators (Weyl, Grübler, Drechsel, Ritthausen).[24]

I shall not discuss Pasteur's great achievements in the study of fermentation processes, except to emphasize the special importance of his discovery of anaerobic microorganisms. That discovery was confirmed by Moritz Traube, who was praised by Pasteur for that contribution. However, there is no mention in any of Pasteur's papers of Traube's views about the nature of fermentation and of the effective agents in that process. For example, in 1878, Traube reiterated views that he had expressed 20 years earlier:

(1) The ferments are not, as Liebig assumed, substances in a state of decomposition, and which can transmit to ordinarily inert substances their chemical action, but are chemical substances related to the albuminoid bodies which, although not yet accessible in pure form, have like all other substances a definite chemical composition and evoke changes in other substances through definite chemical affinities. (2) Schwann's hypothesis (later adopted by Pasteur), according to which fermentations are to be regarded as the expression of vital forces of lower organisms is unsatisfactory. . . . The reverse of Schwann's hypothesis is correct: Ferments are the causes of the most important vital-chemical processes, not only in lower organisms, but in higher organisms as well.[25]

To this may be added that, after years of insistence on the distinction between what he called "organized ferments" or "ferments proprement dites," namely his microorganisms, and "soluble ferments" (such as Schwann's pepsin or Berthelot's invertase), Pasteur confirmed the finding by the Alsatian pharmacist Frédéric Musculus in 1876 of a soluble ferment that cleaves urea, a process that Pasteur had attributed to a microorganism.[26] Pasteur wrote:

The result which we have communicated to the Académie was not and could not have been foreseen. It is the first example of an autonomous organized ferment whose function merges with the function of one of its unorganized products. It is also a new example of a *diastase* [French term for soluble ferment] produced during life and able to modify a substance by the fixation of water, in the same manner as for all the *diastases*.[27]

The next chapter in the story of the development of knowledge about organic chemical isomerism was provided by the man whom Kolbe had derided as a "virtually unknown chemist at a veterinary school"—Jacobus Henricus van't Hoff—and Joseph Achille Le Bel. In September 1874, van't Hoff published a pamphlet (in Dutch) presenting a theory on the extension of chemical formulas into space and on the relation between optical activity and chemical constitution. Two months later, an article appeared by Le Bel on the relations between formulas of organic compounds and the optical activity of their solutions.[28] They arrived at their respective theories independently and, according to van't Hoff, had never talked to each other about them during 1873-1874, when they were both in Wurtz's laboratory in Paris. Although the concept of the asymmetric carbon atom was at the center of both theories, and their names are inseparably linked in the history of chemistry, van't Hoff and Le Bel drew inspiration from different sources. For van't Hoff, it came from the tetrahedral carbon atom of Kekulé and the lactic acid studies of Johannes Wislicenus; for Le Bel it came from the molecular asymmetry of Pasteur.[29] Also, their scientific productivity and professional careers were very different. Although he was five years older than van't Hoff, and outlived him by many years, Le Bel did not gain, nor did he seek, the fame later attained by van't Hoff. The scion of a wealthy Alsatian family, owners of an asphalt factory, Le Bel attended the École Polytechnique from 1865 to 1867, then was an assistant of Liès-Bodard in Strasbourg and of Balard and Wurtz in Paris. In 1875, Le Bel returned to Alsace to take over the management of the family firm (his father had died in 1869), and converted it into a highly profitable refinery of the petroleum extracted from its mines. In 1889, he sold the firm to a German company and moved to Paris. There he acquired a well-equipped private chemical laboratory (with a garden), as well as an elegant house and a pleasant country home for himself and his mother; he never married.[30] During the 1890s, he published numerous organic-chemical papers, and in 1892 was the president of the Paris chemical society, but appears to have been a critic of his French colleagues. In an obituary notice, his friend, William Pope, professor of chemistry at Cambridge, wrote: "His originality of thought, his outspokenness, and his unconventionality, made him somewhat difficult of access, but in congenial society he was a delightful companion, full of knowledge of the world and sparkling with anecdote and caustic wit." [31]

On the other hand, van't Hoff continued to develop his approach to the problem of isomerism, and then achieved great fame as one of the founders of chemical thermodynamics. In 1901, he received the first Nobel Prize for Chemistry and, as professor at Amsterdam (1877-1896) he had many outstanding students. He lived out his years in Berlin. [32] Much more will be said about van't Hoff in the next chapter.

From his youth, when he decided to become a chemist, van't Hoff was also attracted to poetry (especially Byron), philosophy (Comte), and scientific biography. After two years at the polytechnical school in Delft, and a year in Leiden where he mainly studied mathematics, van't Hoff worked during 1872-1873 in Kekulé's laboratory in Bonn, and during 1873-1874 with Wurtz in Paris. Upon his return to Holland, he received his doctoral degree in 1874 at Utrecht, where he then taught physics (1876-1877) at the state veterinary school before his appointment at Amsterdam. In his inaugural address as professor, van't Hoff spoke of the importance of imagination in scientific research. [33]

It has been reported that, in 1904, van't Hoff stated that the stimulus for his formulation of the asymmetric carbon theory came from his reading, in the Utrecht library, of a paper on lactic acids by Johannes Wislicenus. [34] By 1873, at least three compounds having the composition and properties of a hydroxypropionic acid had been identified, one of which (muscle lactic acid) exhibited optical activity, but the preparations were hygoscopic and difficult to purify. In one of his three papers, Wislicenus wrote: "If one admits

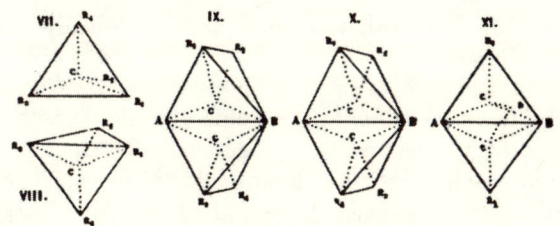

11a. Van't Hoff's tetrahedral diagrams (1874).

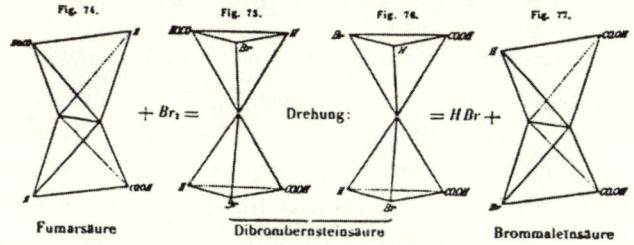

11b. Wislicenus's depiction of the conversion of fumaric acid
to dibromomaleic acid.

Figure 11. Representations of asymmetric carbon atoms

of the possibility of ever distinguishing molecules of the same composition and structurally identical molecules which differ somewhat in their properties, it cannot otherwise be explained than on the assumption that the difference is only based on a different kind of spatial arrangement of the same sequence of the combined atoms"[35] (Fig. 11).

A student and assistant of Wilhelm Heintz in Halle, Wislicenus became a professor in Zürich in 1867, then moved to Würzburg (1872-1885) and succeeded Kolbe in Leipzig.[36] It was from Wislicenus that van't Hoff received the most enthusiastic response to *Chimie dans l'Espace*, copies of which (along with cardboard models) he had sent to numerous leading chemists. After the appearance of the German translation by Herrmann (an associate of Wislicenus), there came the famous blast from Kolbe, while Kekulé, who might have been favorably disposed, was cool (and historically inaccurate):

> Thereby there increases the probability of the hypothesis advanced by Le Bel and further developed by van't Hoff, according to which the four affinities of carbon, already represented as being tetrahedral, are also considered to be spatially tetrahedral. A hypothesis which perhaps does not merit the unqualified praise accorded it by Wislicenus, but in any case much less the harsh ridicule poured on it by Kolbe.[37]

In part, Kekulé's attitude might have been a matter of pique, but still he might have considered the atomic theory only a useful hypothesis, and dismissed the question of the reality of atoms as a metaphysical one.[38] Around 1880, most of the older organic chemists kept alive the separation of the idea of the "chemical atom" from that of the "physical atom," and felt that van't Hoff was attempting to join the two concepts. Subsequent experimental work in the laboratories of van't Hoff and Wislicenus, as well as those of Arthur Hantzsch, Victor Meyer, and Alfred Werner, showed, however, that in the assignment of molecular configuration, chemical methods (reaction conditions, purification procedures) can be usefully complemented by physical methods.[39]

The climate of opinion among organic chemists shifted markedly in favor of the van't Hoff theory in 1887 with the publication of a long paper by Wislicenus summarizing its utility in the elucidation of the structure of maleic acid, fumaric acid, and other unsaturated compounds.[40] During the last quarter of the nineteenth century, the main interest of leading German organic chemists—Adolf Baeyer, Emil Fischer, Emil Erlenmeyer, Rudolf Fittig, Victor Meyer (who introduced the term "stereochemistry")—was in the synthesis of natural products, dyes, and drugs. Although there were disagreements about some of the conclusions drawn by van't Hoff and Wislicenus, by the 1890s the general principles of their theory of stereochemical isomerism had largely been accepted in Germany, with its brilliant application to the constitution and to the synthesis of the monosaccharides by Emil Fischer[41] (Fig. 12b). Two parallel developments broadened the scope of studies on the isomerism of organic compounds. One was the explanation of the phenomenon that came to be called "tautomerism," as in the inter-conversion of the keto- and enol-forms of ethyl acetoacetate.[42] The other was the recognition that cyclohexane, the fully hydrogenated derivative of benzene, does not lie in a single plane. Baeyer had assumed that, in line with his "strain" theory, cyclohexane is fully planar, but in 1890 Hermann Sachse described two strain-free "configurations" (they were later termed "conformations"), one a rigid "chair" form, the other a flexible "boat" form. Both the strain theory and conformational analysis formed active areas of investigation during the twentieth century.[43] Among the other extensions of van't Hoff's theory was the important study by Arthur Hantzsch and Alfred Werner of the isomerism of nitrogen compounds, with the later preparation, by Werner (of whom more shortly) of an optically active compound containing no carbon atoms[44] (Fig.12a). Moreover, in 1896 Victor Meyer described the phenomenon of "steric hindrance" in the rotation of chemical groups about a single carbon-carbon bond, and Paul Walden reported that an "optical inversion" could be effected in some optically active substances.

Apart from the cool reception given the theories of van't Hoff and Le Bel by Kekulé and Pasteur, during the 1880s some organic chemists, notably Adolf Claus and Wilhelm Lossen rejected van't Hoff's theory on the ground that it violated the laws of physics. The most cogent criticism came from the American chemist Arthur Michael.[45]

Of particular interest is the attempt, during the 1890s, of Le Bel to emphasize the difference between his formulation of the theory of the asymmetric carbon atom from that of van't Hoff:

> I used the greatest efforts in all my explanations to abstain from basing my ideas on the preliminary hypothesis that the compounds of the formula CR^4 have the shape of a regular tetrahedron. It has happened that very many scientists who

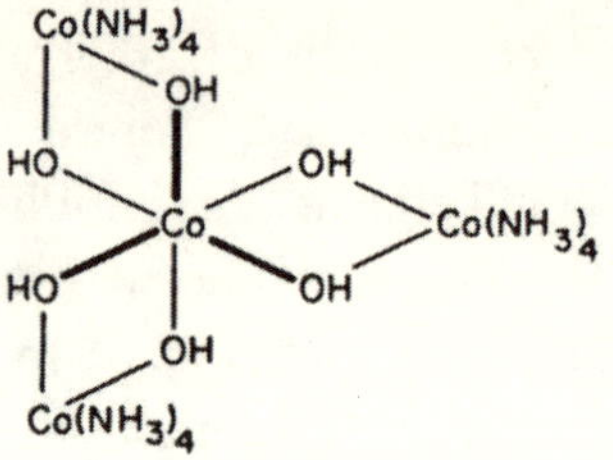

12a. An optically-active coordination compound not containing carbon (Werner, 1914).

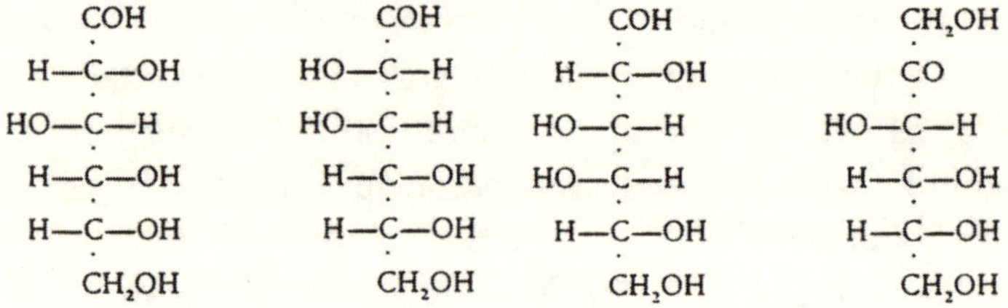

d-Glucose *d*-Mannose *d*-Galactose *d*-Fructose

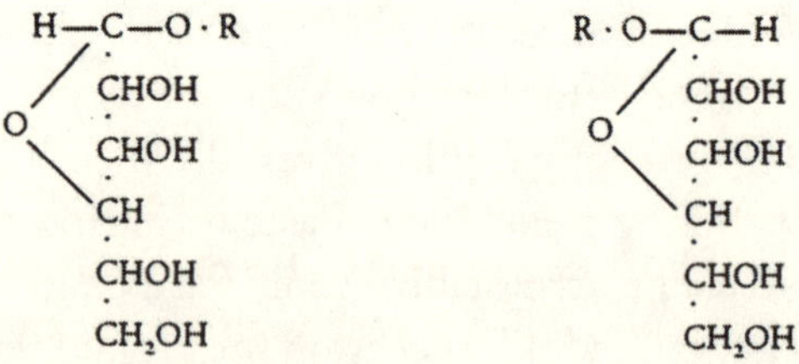

α-Methylglucoside β-Methylglucoside

12b. Emil Fischer's projection formulas for sugars and glucosides (1894).

Figure 12b. Fischer projection formulas

wrote about my article most favorably in other respects, did not turn their attention to the fundamental difference, which exists between my starting point and the starting point of M. van't Hoff in his analogous work published at the same time in Utrecht.[46]

Indeed, in its present-day form, stereochemistry is not defined either by the presence of an asymmetric carbon or nitrogen atom, or of optical activity. As Vladimir Prelog put it:

Enantiomers are optically active but the enantiomerism is a geometrical feature; the optical activity is only a consequence of the symmetry, as are the dipole moments of compounds without a center of symmetry. To determine the scope and limits of stereochemistry it is necessary to trace it to its source and that is chirality.[47]

As for Pasteur's conviction that the optical activity of organic compounds was a consequence of their creation by living organisms, and his failure to induce optical activity artificially (by magnetism), that idea reappeared in 1898 in a lecture by the chemist Francis Japp, who asserted that "a directive vital force came into play the moment of organic creation of 'a force of precisely the same character as that which enables the intelligent operator, by the exercise of his Will, to select one crystallized enantiomorph and reject its asymmetric opposite.'"[48]

As was noted earlier, Alfred Werner must be added to our list of notable Alsatian chemists—Wurtz, Gerhardt, Le Bel. He began chemical work in his native Mulhouse with the dye chemist Emilio Noelting, and in 1886 went to Zürich, where he received his Ph.D. degree in 1890 for the study of isomeric nitrogen compounds. After a year with Berthelot in Paris, Werner returned to Zürich, and became associate professor (1893) and full professor (1895) at the university.[49]

In his *Habilitationsschrift* (1890), Werner advanced the view that it was necessary to replace the definition of valence, developed by Kekulè, van't Hoff, and Wislicenus with a limited number of single forces acting toward a regular tetrahedron, by the assumption that "affinity is an attractive force acting equally from the center of the atom toward all parts of its spherical surface. It necessarily follows from this concept of affinity that separate valence units do not exist. Valence signifies an empirically determined numerical relationship independent of valence units." [50] Then followed a memorable series of papers by Werner on the constitution of the long-known "metal-ammines," [51] which Kekulé had denoted as "molecular compounds" outside the scope of his valence theory. In 1893, Werner introduced the terms *Hauptvalenz* (primary valence) and *Nebenvalenz* (secondary valence), and proposed that every metal has a definite number of secondary valences ("coordination number") spatially directed around a central metal atom acting as an anion.

Werner applied this theory to the study of the constitution of the compound whose composition is $CoCl_3(NH_3)_6$, for which various structures had been proposed by previous investigators, and he offered convincing evidence for the view that the six ammonia molecules are coordinated with the cobalt atom by secondary valence to form a 3+ ion (the primary valence of the metal is 3+), which is neutralized by the three chloride ions. In this case, the maximum number of atoms or molecules that can be attached to the metal is six. Werner proposed that the ammonia molecules lie at the corners of a regular octahedron, with the cobalt at the center. Many other compounds of this type were known, and Werner added to their number, in which one or more of the ammonia molecules had been replaced by other molecules such as water, ethylene diamine ($H_2NCH_2CH_2NH_2$), or nitro groups, or in which the primary valence of the cobalt was 2+. Moreover, he explained the constitution of coordination compounds involving other divalent metals, for example Zn^{2+}, in the same manner. [52] One of the predictions stemming from the formulation of the octahedral structure was that some of the compounds with ethylene diamine as ligands should exist as two diastereoisomers. Werner's group expended much effort over a period of about 9 years before one of them was resolved in the form of salts of the dextrorotary camphorsulfonate. [53] To eliminate the possibility that the optical activity was due to the presence of carbon, Werner demonstrated the resolution of a carbon-free octahedral compound. He also showed that for elements with a coordination number of four, where the arrangement of the ligands may be planar or tetrahedral, two geometric isomers are possible.

Werner's coordination theory represented a rejection of the views of Christian Wilhelm Blomstrand and Sophus Mads Jørgensen, and during 1893-1899 Werner and Jørgensen conducted a lively debate. Any skepticism about the merits of Werner's theory was set aside after the announcement of the successful resolution of a coordination compound to produce optically active isomers. [54] The predictive power of Werner's theory became even more evident during the early 1920s, when J.N. Brønsted and T.M. Lowry developed the concept of acids as proton donors and bases as proton acceptors, an idea advanced by Werner in 1907, and when the X-ray crystallographic diffraction studies of R.W.G. Wyckoff and of R.G. Dickinson on platinum and palladium coordination compounds showed them to be octahedral in structure.

The importance Pasteur, Japp, and others attached to the role of living organisms in the formation of optically active compounds became more explicit, and less mystical, when during the 1890s Emil Fischer showed that α-methyl glucoside is hydrolyzed by a preparation of the enzyme invertase (studied by Berthelot) but not by a preparation of emulsin (discovered by Wöhler and Liebig), whereas β-methyl glucoside is cleaved by emulsin but not by invertase.[55] The advances made during the twentieth century in the development of methods to purify enzymes made the reliable study of their chiral discrimination and their mechanism of catalytic action possible. The knowledge so gained has illuminated the findings of pharmacologists on the differential effect of enantiomeric forms of drugs used in medical practice. Apart from the fundamental importance of the development of new techniques for the asymmetric synthesis of organic compounds, this area of modern chemistry has become crucial in the design and development of new medicinal agents. Moreover, stereochemical theory has figured prominently in speculations about the origin of life.[56]

To conclude this chapter, I return to the development of the art of organic synthesis, with special reference to its utility in establishing the validity of a particular formula for the molecular structure of an organic compound, and in preparing potentially useful derivatives of that compound. Before the acceptance, chiefly in Germany, of the theories of valence, structure, and stereochemistry, the objectives of organic synthesis were rather different. By 1860, Berzelius's original definition of organic compounds as those produced by living organisms had been replaced by the view that organic chemistry deals with carbon compounds, and not with the "organized" biological substances. The many synthetic achievements before about 1860, such as Wöhler's synthesis of urea (and, like it, frequently a matter of accidental discovery), were seen as steps to unify organic and inorganic chemistry, to justify drawing analogies from the inorganic realm in formulating the constitution of organic compounds according to the radical or type theories, and to explain the existence of isomeric substances.[57] What remained after these theories were discarded were the valuable preparative reactions introduced by men such as Strecker, Frankland, Hofmann, and Kolbe. The French approval of the claim of Berthelot, an anti-atomist, to have "banished life from chemistry" by the pyrogenic synthesis of small molecules such as acetylene, was not shared by his chemical contemporaries in Germany.

During 1880-1914, the German chemists engaged in synthetic work greatly outnumbered those in France and England, and two of them—Adolf Baeyer and his pupil Emil Fischer—outshone all the rest. Only a few of the others need be mentioned: Rudolph Fittig, Ludwig Claisen, Theodor Curtius, Johannes Thiele, Otto Wallach, Richard Willstätter, and Heinrich Wieland. In France, there were Charles Friedel, Victor Grignard, and Paul Sabatier, and in England the William Henry Perkins's (father and son). The collaboration of many of the German university chemists with a burgeoning industry engaged in the large-scale manufacture of dyes and drugs was an important factor in the domination of the world market by companies such as Merck, Bayer, Höchst, and Schering. Moreover, some of these chemists—notably Fischer, Willstätter, and Wieland—showed the way in the application of synthetic chemistry to the study of "organized" biochemical products such as the proteins, enzymes, and plant pigments.

As professor in the Berlin technical school (1860-1872), Strassburg (1872-1875), and Munich (1877-1915), Baeyer's personal research dealt successively with a variety of projects, among them the constitution of uric acid, the structure of indigo, the condensation of aldehydes and ketones with aromatic compounds, the synthesis of phthalein and triphenylmethane dyes, and the structure of hydrogenated benzene derivatives, oligoacetylenes, terpenes, and organic peroxides.[58] A devoted pupil of Kekulé, whose theories he eagerly embraced, Baeyer stated that, in contrast to his teacher who, as was noted earlier, was only interested in substances if they fit into his theories, "I do not make the experiment in order to see whether my views are or are not correct, but in order to find out what the behaviour of the substance is under a variety of conditions. It is for this reason that I attach little importance to theories."[59] As is common among self-proclaimed empicirists, Baeyer staunchly defended his strain theory, and offered speculations about the chemical pathways in plant metabolism.[60]

According to his contemporaries in Berlin, the 30 year-old Baeyer was a man of fine presence, and some of the qualities he displayed in later years—"a steadfastness of purpose and determination to allow nothing to interfere with his duties as a teacher or with the course and development of his investigations. He made it quite clear to the University authorities that he would have nothing to do with general University politics, and he always refused to serve on committees or attend meetings unless they were concerned with the affairs of his department."[61] His style of personal research became legend. He depended largely on his own assiduous test-tube experimentation, and even when mechanical devices (for example, shaking machines) were available, he preferred stirring rods. Also, from his days in Berlin, he was exceptional among German chemistry professors in his generosity toward his junior research associates. For example, in 1868 his associate, Carl Graebe, and a student, Carl Liebermann, used Baeyer's zinc dust distillation method to determine the parent substance of natural alizarin (a widely used textile dye). It turned out to be anthracene, thus leading to the synthesis of alizarin (dihydroxyanthraquinone) and to its large-scale manufacture. Baeyer later repeatedly hailed Graebe's success, but made no public claim for his part in that achievement. During his 40 years in Munich, Baeyer had a succession of postdoctoral *Privatassistenten*, usually drawn from his former Ph.D. students, but only 45 doctoral dissertations (out of 395) were based on work within the areas of Baeyer's personal research interest. As William Henry Perkin, Jr., who worked in Baeyer's institute during the 1880s described it:

Perhaps for the reason that he was accustomed to do all the experimental work himself, Baeyer had little inclination to work with others, and the titles of his papers show that whilst he frequently published with one or another of his assistants . . . the actual number of his coworkers was relatively small. This is all the more surprising when it is remembered that the laboratory was always overcrowded with the most promising material from all parts of the world, and that every newcomer would have considered it a great honour to have been allowed to work with the head of the laboratory. Neither did Baeyer often suggest subjects for the researches which so many of the younger men were carrying out for their dissertations for the Ph.D. degree. When the time came to do original work, these young researchers were usually handed over to one of the many Privatdozenten attached to the laboratory, and it was the duty of these senior men, who were often men of great experience, to suggest the theme for investigation and to superintend the work and help to bring it to a satisfactory conclusion. This plan worked well, and gradually there rose up a great school of research which has rarely if ever been equalled. . . . There was a feeling in the laboratory that no one was of any account who did not research, and, moreover, the position of each researcher and the esteem in which he was held depended on the quality of the work he was engaged in. This was the atmosphere which produced the greatest chemists of the day and weeded out those who were of no account. It was only necessary for the commanding figure of Baeyer should stroll through the research laboratories each day and for him to chat with the various workers, criticize their results, and admire their preparations, to make it out of the question for anyone to forget for a moment that research was the only thing that really mattered.[62]

A slightly different appraisal was provided by Hans Rupe, who had also been one of Baeyer's Ph.D. students and *Privatassistenten* around 1890:

> During my time in Munich the *Praktikanten* in the organic section did not come face to face with the "old man" often, and the ones in the inorganic section on the floor above even less frequently. When he occasionally strode through the halls, erect, every inch a king, in his indigo blue long cutaway, the hat on his head, with a stern countenance, there swept through the room a slight wave of uneasiness, everyone crouched behind his work-bench, and sociable conversations were abruptly interrupted. He was considered to be the "Stiff Prussian," unapproachable, strict, pitiless. Whoever had ears to hear, however, knew that at bottom Baeyer was a very kindly soft-hearted person who quietly did much good and eased many difficulties. He belonged to the few scholars and chemists who acknowledged wholly and ungrudgingly the merits of others. The colleagues in the laboratory thought "he can afford it"; yes, but many others could also afford it, but did not do it.[63]

It should be added that, at a time when anti-Semitism was endemic in German academic life, Baeyer gave encouragement to his pupils Victor Meyer, Eugen Bamberger, and Richard Willstätter, and was a close friend of the great dye chemist Heinrich Caro. Baeyer's attitude was no doubt influenced by the fact that his mother (née Hitzig) had Jewish forebears. This did not go unnoticed by the Nazis; his sons Hans (professor of medicine at Heidelberg) and Otto (professor of physics at Berlin) were both dismissed from their university posts during the 1930s "auf Grund rassischer Abstammung."

Among Baeyer's scientific progeny, Emil Fischer overshadowed all the others in scientific achievement and public prestige.[64] Although some of Baeyer's personal qualities were reflected in Fischer's demeanor, his style of research and of leadership differed markedly from those of his teacher, especially after his ascension to the professorship in Berlin in 1892. As a Ph.D. student and Privatdozent (1878) in Strassburg, and as an associate professor (1879) in Munich, Fischer first worked on dyes (fluorescein, rosaniline), but then shifted to the study of derivatives of hydrazine. In extending the earlier work of Adolf Strecker on diazonium compounds, Fischer discovered phenylhydrazine ($C_6H_5NHNH_2$), which, in his hands, provided an entry into the elucidation of the structure of the sugars. Fischer's prolonged exposure to phenylhydrazine and other toxic agents led to his life-long painful ill-health, which must be taken into account in a consideration of his style of research. In 1884, he found that the reaction of glucose and fructose (previously shown by Heinrich Kiliani to be straight-chain compounds) gave the same crystalline derivative (an "osazone"), and that an isomeric osazone was produced with galactose. During the succeeding 15 years, Fischer's research group established the structure and stereochemistry of many sugars, and effected their synthesis by means of methods that were elegant and lasting. Apart from the ingenuity and skill displayed in this achievement, which marks the high point of his career, Fischer's use and development of the van't Hoff-Le Bel concept of the asymmetric carbon atom bespeaks the theoretical insight he brought to the problem.[65] In connection with his stereochemical studies, he made an important contribution to the explanation of the so-called Walden inversion of configuration. The work on carbohydrates also marks Fischer's entry into biochemical research, through his studies on the fermentation of the sugars and the enzymatic cleavage of glycosides, leading to his famous lock-and key analogy of the specificity of enzyme action. And, as a by-product of his work on derivatives of phenylhydrazine, Fischer discovered a route for the synthesis of indoles.

Before beginning his work on sugars, Fischer undertook the study of caffeine because he questioned the validity of the proposal (1875) of Ludwig Medicus that its structure is closely related to that of uric acid, whose constitution Baeyer had not succeeded in determining during the 1860s. In a series of elegant synthetic experiments between 1881 and 1898, Fischer provided definitive evidence for the correctness of Medicus's idea, and showed that caffeine and uric acid, as well as xanthine and guanine, are derivatives of a parent substance that Fischer named "purine." During the 1880s, largely through the work of Albrecht Kossel, it became known that the purines guanine and adenine are constituents of nucleic acids. Fischer's Nobel Prize (1902) was awarded for his work on sugars and purines.

Fischer's public fame rose sharply around 1905, when it was expected that he would shortly effect the laboratory synthesis of a protein. He entered this field in 1899, with an ambitious program that began with the laboratory synthesis of racemic amino acids, their resolution to yield the enantiomers corresponding to the forms obtained upon the acid hydrolysis of proteins, the analysis of such hydrolysates for their content of the individual amino acids, and the synthesis of linear assemblies of amino acid units to form what Fischer named "peptides." This peptide theory of protein structure, which was also advanced by the physiological chemist Franz Hofmeister in 1902, led Fischer to expect that, with increasing length of the synthetic peptide chains, the products would first have the properties of peptones and albumoses (products of the partial degradation of proteins by enzymes or acids), and that eventually a chain length would be reached (perhaps 20 amino acid units) corresponding to the size of natural proteins. The synthetic method he devised for the sequential addition of amino acid units to the growing chain involved the use of reagents such as chloroacetylchloride ($ClCH_2COCl$), which combines with the amino group of the chain to form the chloroacetyl-peptide ($ClCH_2CO$-$NHCH(R)CO$-). In 1905, Fischer wrote to Baeyer:

> On January 6th I will present a lecture at the Chemical Society summarizing my work on amino acids, polypeptides and proteins, and then early next year I will publish the collected papers in the form of a book. The material has grown splendidly and there is much detail in it. I have also prepared the first crystalline hexapeptide and hope to obtain a matching octapeptide before Christmas. Then we should be close to the albumoses. . . . My entire yearning is directed toward the synthesis of the first synthetic enzyme. If its preparation falls into my lap with the synthesis of a natural protein material, I will consider my mission fulfilled.

A few months later, he wrote Baeyer: "The synthesis of the polypeptides is advancing briskly. I have recently made the first decapeptide and will now try to reach the eicosapeptide, whereupon one should be midway in the protein group."[66]

In 1907, Fischer claimed that "*l*-leucyl-triglycyl-tyrosine prepared artificially has all the properties of the albumoses. These observations are of importance in casting doubt on the view which formerly prevailed that, being intermediate products between proteins and peptones, the albumoses are substances of considerable complexity." He also asserted that "the synthesis of the higher terms [peptides] has been restricted hitherto to the combinations of glycine, alanine, and leucine; there is not a shadow of doubt, however, that all the remaining amino acids could be associated in complicated systems with the aid of our present methods."[67]

Fischer's confidence in the power of his method proved to be short-lived, for it was not only cumbersome and costly, but also unsuitable for the preparation of polypeptides containing more complex amino acids, for example lysine or glutamic acid. The enormous effort of his assistants and students produced much less than he had hoped for, and after 1910 there were no further experimental publications on peptide synthesis from his institute. He continued to insist, however, that the molecular weight of proteins did not exceed 5,000, and dismissed the physical-chemical evidence for larger molecular weights. As subsequent work showed, Fischer's synthetic strategy was sound, but new reagents were needed to effect the coupling reactions and the protection of reactive groups in more complex amino acids. During the last decade of his life, Fischer returned to carbohydrate and purine chemistry, attempted to enter the nucleic acid field through the synthesis of nucleosides, and made important contributions to the study of tanning agents. This work was done by his postdoctoral assistants, because of Fischer's involvement in the organization of the Kaiser-Wilhelm institutes and, during World War I, his service in the chemical aspects of the war effort.[68] It must be noted that, in addition to his continuing ill health, Fischer was beset by family tragedies. His wife Agnes died in 1895, and two of his three sons died during World War I.

In contrast to Baeyer's policy in Munich, Fischer considered the organic-chemical part of his Berlin institute to be largely a laboratory for the investigation of problems of immediate interest to him. Whereas in Würzburg he encouraged the independent efforts of some of his junior associates, notably Ludwig Knorr, in Berlin his attitude was less generous in that regard. One of the men who resisted, and who survived, was Carl Harries, a holdover from the Hofmann regime, probably because he had married the daughter of the noted industrialist Werner von Siemens. For a beginner in Fischer's laboratory, his appearance must have been awesome. A few days after Otto Diels began work there in 1896, "a tall, very erect man with a pince-nez and black beard strode through the hall, followed by a laboratory servant. Diels was struck by the fact that he wore a black hat and a blue tunic. Diels asked his neighbor who the man was, and received the amazed reply: 'Don't you know? Why, that was Fischer.'" [69] Another report suggests something of the atmosphere of the laboratory:

> With a stern eye he inspected the laboratory workers, who reported to him the progress of their experiments. Fearsome was his *Flügelschlagen* [flapping of wings], without further comment, for the poor wretch if something had gone thoroughly wrong. Only rarely did the chief sit on a stool and conduct a brief private conversation. Then it was even permissible to laugh. [70]

An American guest worker, the physician James Bryan Herrick, offered a somewhat different report: "He was modest, kindly, always the gentleman. Twice a day he made the rounds, moving quietly from desk to desk inspecting the work, always seeming interested, criticizing, helpfully suggesting. He had the faculty of seeing quickly where one's trouble lay. So gentle in manner was he that one scarcely realized that he was a good executive commanding officer." [71]

Another aspect of Fischer's style of leadership was noted by his last chief research assistant, Max Bergmann, who wrote that Fischer was reticent to his co-workers

> when he gave them instructions for the conduct of experiments or himself did laboratory work in their presence. Then, an indication of the purpose and goal and expected outcome of the experiment was either not given or stated very incompletely. The explanation of this behavior may be found in a printed guide for the conduct of scientific experiments, which Fischer regularly presented to the older students of his institute and to his own assistants. One sentence was: "You are urgently warned against allowing yourself to

be influenced in any way by theories or by other preconceived notions in the observation of phenomena, the performance of analyses and other determinations."[72]

Like his teacher Baeyer, Fischer discouraged chemical speculation by his research assistants, but allowed himself to have
"theories and preconceived notions" of his own. He also imi-
tated Baeyer in his reluctance to use physical instruments
in chemical work. For example, he considered that, for the
detection of some substances, the sense of smell was more
valuable than spectrum analysis.[73]

To these facets of life in Fischer's institute must be
added the matter of secrecy, not an uncommon phenomenon in organic-chemical laboratories. Hans Thacher Clarke,
an Englishman (later professor of biochemistry at Columbia University) who worked at the Berlin institute during
1911-1913, wrote that upon his arrival there he was advised

> not to ask the other members of the laboratory what they
> were doing. This was so contrary to British tradition that
> I was interested to find out the reason; it appeared that
> most of the chemists who were working on topics of their
> own were retained as consultants by one or another of the
> German manufacturing firms, which had priority on any
> patentable discoveries made by the individuals concerned.
> This system appeared to me, and still does, as being at vari-
> ance with the prime function of an academic laboratory.[74]

Fischer himself received royalties from patents on dyes (rosaniline) and drugs (veronal, sajodin), and Carl Duisberg, the
head of the Bayer company, was one of his closest friends.

Finally, I must note that although Fischer's personal
researches in Würzburg and Berlin outshone those of Baeyer
in Strassburg and Munich, Baeyer's laboratory in Munich
served as a more fertile breeding ground for the next generation of leading German organic chemists. I list only ten

of them, roughly in the order of the start of their work in that laboratory: Wilhelm Königs,[75] Hans von Pechmann,[76] Ludwig Claisen,[77] Eugen Bamberger,[78] Theodor Curtius,[79] Johannes Thiele,[80] Richard Willstätter,[81] Otto Dimroth,[82] Heinrich Wieland,[83] Kurt Hans Meyer.[84]

Among the many later notables from abroad were (from England) William Henry Perkin, Jr.,[85] and (from the United States) Moses Gomberg and William Albert Noyes.[86] The subsequent development of the art of organic synthesis in Germany, England, and the United States (and other nations) by the scientific progeny of these men is too large a story to be told adequately in this book.[87]

Chapter Eight

FORCES, EQUILIBRIA, AND RATES

ॐ

I HAVE REFERRED to the distinction made by some nine-teenth-century organic chemists (for example, Butlerov) between the concepts of chemical atoms and of physical atoms. The idea of the chemical atom, with a characteristic relative weight, explicit in Dalton's theory, and the principles of definite and multiple proportions, proved to be fruitful in the formulation of successive theories (radicals, types, residues) of molecular constitution and interaction. The main challenges were seen to be in the accurate determination of atomic weights, in the explanation of affinity and isomerism, and in the classification of the rapidly growing number of known and newly discovered natural substances and of the new substances artificially produced in the laboratory. Except for those chemists who drew inspiration from the Avogadro-Ampère molecular theory, and from crystallography (Ampère, Gaudin, Baudrimont, Laurent), polarimetry (Pasteur, Le Bel), or spectroscopy (Stokes), the physical instruments in general use were improved versions of the eighteenth-century balances, thermometers, calorimeters, and gasometers. Few of the leading contributors to the development of organic chemistry were well versed in the physics and mathematics of their time, and efforts by men such as Cayley, Sylvester, and Brodie to mathematize chemistry bore relatively little fruit.

Moreover, before about 1870, many of the leading chemists were skeptical about the real existence of the atoms of Dalton's theory, and (as in the case of Wollaston) preferred to use the concept of equivalents or (as in the case of Davy) to consider variants of the concept offered by eighteenth-century natural philosopher Rudjer Bošković, who invoked the occurrence of alternating attractive and repulsive forces between elements of matter, with points of stable equilibrium perceptible as material units in crystals.[1] The atomic debates took somewhat different forms in England, Germany, and France, and while some chemical skeptics were converted to atomicity by the stereochemistry of van't Hoff, Le Bel, and Wislicenus, the debate was resumed at the turn of the century with the advocacy of "energetics" by the anti-atomistic thermodynamicists Wilhelm Ostwald and Pierre Duhem.[2]

The concept of the physical atom, as seen at mid-century by some French, English, or German organic chemists, was that of a unit subject to the mathematical laws of mechanics as formulated by Descartes, Newton, or Leibniz. During the first decades of the nineteenth century, the phenomena of heat, light, electricity, and magnetism, once topics within the orbit of eighteenth-century chemists, had become the province of experimental and mathematical investigators whom William Whewell later labeled "physicists" (among "scientists") such as Arago, Biot, and Fresnel in the study of light, or Ampére and Faraday in the study of electricity, or the mathematician Fourier and the engineers Carnot and Clapeyron in the study of heat. The eighteenth-century ties to chemistry remained unbroken, however, and there was speculation about the possibility that light, heat, and electricity are simply different manifestations of the same fundamental principle, and usually denoted as an "imponderable fluid." From these beginnings, new insights into the concepts of chemical energy, the equilibria and kinetics of chemical reactions developed, as did, with the discovery of the electron by J. J. Thomson, the nature of valency and the chemical bond. This interplay of physical and chemical thought and experiment brought forth the emergence of physical (or theoretical) chemistry as an independent discipline, with a primary interest in chemical thermodynamics, electrochemistry, and the theory of solutions.[3]

The concept of chemical energy did not emerge clearly until the middle of the nineteenth century. Before then, the words force or power (German *Kraft*) were used to denote the capacity to produce physical motion or chemical change.[4] Although the idea that heat is a form of motion had been advanced earlier by Bacon, Newton, and Locke, during the eighteenth century, heat was considered to be an imponderable material substance by Black, who distinguished between heat and temperature, and Lavoisier, who included his "calorique" (along with light) in his table of "simple substances."[5]

The caloric theory of heat still held sway in 1824, when Sadi Carnot published his *Réflexions sur la Puissance Motrice du Feu*, in which he presented an analysis of the operation of an ideal form of Watt's steam engine as a series of reversible steps in a cyclical process involving the expansion and compression of a gas between two temperatures. He likened the action of caloric in driving a heat engine to that of falling water in driving a water wheel, and thought that the flow of heat from a hot body to a cold body could produce work without the loss of heat. He recognized, however, that in the operation of a non-ideal engine, only a fraction of the heat is convertible to work, and is given by the expression $(T_2 - T_1)/T_2$. Thus, the work done by a steam engine per amount of fuel is increased by larger differences in the temperature, and the efficiency is decreased by operating the engine at a higher initial temperature. Carnot also calculated that 1.12 units of work were yielded when 1,000 units of heat went from 100° to 99°. A brilliant Polytechnicien and military engineer, Carnot fell victim to cholera in 1832, at the age of 36. Two years later, another Polytechnicien, the engineer Benoit Pierre Emile Clapeyron, rescued Carnot's memoir from obscurity, and developed a mathematical and graphical version of Carnot's cycle. It should be noted that Carnot was an isolated investigator, outside the Parisian community of physicists, and that he derived stimulus from the work of the chemists Nicolas Clément and Charles Bernard Desormes.[6]

By about 1850, the caloric theory had been replaced by the concept of heat as a mode of the motion of the particles constituting matter. Also, although Carnot's idea, in his 1824 paper, of the conversion of heat to work had undergone drastic revision, the Clapeyron equation was used by later investigators as the starting point in the formulation of new theories. The chief figures in this development, which included the formulation of the principle of the conservation of energy (the "first law"), were James Joule, William Thomson (later Lord Kelvin), and Rudolf Clausius. They were preceded during the 1830s by Carl Friedrich Mohr and Ludvig August Colding and, during the 1840s, the physician Robert Julius Mayer and the physiologist Hermann Helmholtz, whose role, in my opinion, has been unduly magnified by some historians.

James Prescott Joule came from a Manchester family that owned a large and profitable brewery. He was a sickly child and derived his formal education from tutors, among them the aged John Dalton, who taught him mathematics and elementary chemistry (in his later life, Joule succeeded Dalton as president of the Manchester Literary and Philosophical Society). Young Joule began his experimental work at the brewery, where he helped in its management; his subsequent research was conducted in a laboratory in his parental home, and then in his own private residences.[7]

Manchester was one of the chief centers of the British Industrial Revolution, with extensive use of steam power in its factories.[8] During the 1840s, the intellectual atmosphere was rather different from that in London, where the Royal Society tended to be dismissive of the scientific efforts of the engineers and amateurs who sought to devise more efficient methods of generating power.[9] Their research reports were frequently rejected by the Royal Society for publication in its Proceedings, and appeared in journals such as the *Philosophical Magazine*, or became known through oral presentations at public meetings such as those of the British Association for the Advancement of Science. This was the case for the engineer William Sturgeon, who stimulated Joule's early work, and for Joule himself. It was at one of the meetings of the British Association that he gained the attention of William Thomson (later Lord Kelvin).

Joule's research reports gave evidence of the skill and ingenuity with which he planned and performed his experiments, and of his persistence in striving for the most accurate results. He first attempted to devise an electromagnetic engine that was more efficient than the best steam engines of his time, but soon realized that these efforts were fruitless. He set about to investigate the quantitative relation of electricity to heat. He found in 1843 that the heat generated in an electric circuit is proportional to the square of the current and to the resistance of the wire, and later asserted that "electricity may be regarded as a grand agent for carrying, arranging, and *converting* chemical heat."[10] He then took up the problem of the conversion of heat to mechanical work. Among his several experimental approaches was the use of an apparatus in which the heat (measured in a calorimeter) was generated by the rotation in water (or another fluid) of a paddle wheel operated by movement of weights on pulleys (Fig. 13). Another method involved the measurement of the heat generated by passing

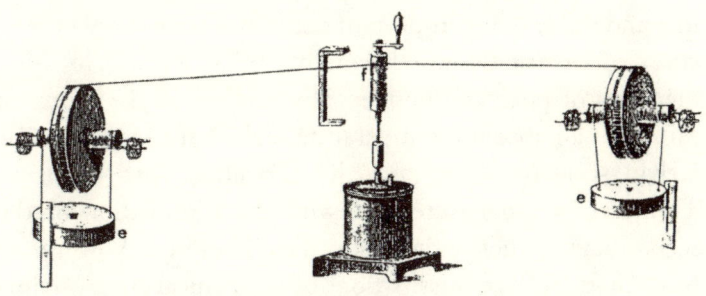

Figure 13. Joule's apparatus for the measurement of the
mechanical equivalent of heat (1845).

water through narrow tubes. Joule's values for the mechanical
equivalent of heat, as obtained by different methods, ranged
from 774 to 890 ft-lbs to raise the temperature of 1 lb of water
by 1°F (later to be named the British thermal unit, BTU). The
lower value corresponds to 424.5 kilogram-meters to raise the
temperature of 1 kg of water 1°C (the officially accepted value
is 427 kg-m/kcal).[11]

By 1845-1847, Joule had been fully converted to the dynami-
cal theory of heat, and the principle of the conservation of
energy. In 1847, his influential fellow Mancunian, Lyon Playfair, with whom Joule collaborated for a few years, urged him
to apply for a professorship at the University of St. Andrew's,
but it appears that, after eliciting some recommendations,
Joule decided to remain in Manchester. In 1847, the wider recognition of Joule's contributions led not only to his friendship
with Thomson, but also to the priority claim of Robert Julius
Mayer and to the adverse criticisms of Joule's experimental
work by Hermann Helmholtz.

After graduating from Cambridge in 1845, Thomson spent
a year with Regnault in Paris, where he became acquainted
with the writings of Carnot and Clapeyron. In the following year (at age 22), he was appointed professor of natural
philosophy at Glasgow, where he remained the rest of his

life, and made many important contributions not only to the study of electricity and thermodynamics but also to other branches of pure and applied physics.[12] In 1848, Thomson proposed an absolute temperature scale, later known as the Kelvin scale ($0°$ C = $273.1°$ K). Together with Clausius, Thomson is usually credited with the formulation of the second law of thermodynamics. For Thomson, however, it was a matter of the "dissipation" of mechanical energy, while Clausius emphasized the partition of the heat of a material body into a portion available for work and a portion that he later termed the "entropy" associated with the "bound energy" and unavailable for mechanical work.

As for Julius Robert Mayer, in his account of the line of his thought about the relationship between heat and work, Mayer told of his observation, while a ship's physician in the tropics, that the venous blood of ailing and inactive sailors, upon being subjected to blood-letting (a standard medical practice), had the bright red color of arterial blood, but in healthy, working sailors the venous blood was dark. Drawing on Lavoisier's description of animal respiration as an oxidative combustion, Mayer concluded that in the tropics, less blood oxygen was used up than in a colder climate or upon heavy labor, and he claimed that, during the return voyage to Germany, he had studied the physics of heat, and had concluded that there was an equivalence between the amount of work done and the heat of respiratory combustion generated in the body. In 1842, Mayer published in Liebig's *Annalen* a paper in which he offered some eccentric views about the nature of "force," and used a method described by Mohr to calculate the mechanical equivalent of heat from calorimetric data in the literature; Mayer's value was 365 kg-m/calorie to raise the temperature of one kg of water $1°$ C. Three years later, he offered his opinions on the relation of work to the heat generated in animal metabolism. Although he drew on the physiological speculations in the 1842 edition of Liebig's *Animal Chemistry*, Mayer disputed Liebig's view that oxidation of muscle protein was connected to muscular work, and insisted that all of the respiratory combustion is effected in the blood.[13] In 1848, Mayer published a theory of the origin of the heat and light of the sun, as being caused by the impact of meteors, and also used the pages of the *Comptes Rendus de l'Académie des Sciences* (16 October) to affirm his priority in the determination of the mechanical equivalent of heat. Joule's response appeared in the issue for 22 January, 1849. He later wrote Thomson that "I . . . shall be quite content to leave Mr. Mayer in the enjoyment of having predicted the law of equivalence. But it would certainly be absurd to say that he has established it. I have not pursued the controversy further because the facts are before the scientific world and shall be perfectly satisfied with its verdict whatever it might be."[14] It has been reported that the indifference of German scientists to Mayer's writings, as well as family troubles, caused Mayer to suffer, in 1850, a mental breakdown and an attempted suicide. Indeed, during the 1850s, his chief defender was John Tyndall, then head of the Royal Institution and admirer of German scientists.

In Hermann Helmholtz's famous 1847 paper on the conservation of force, no mention was made of Mayer, nor of Mohr and Colding, and the several references to Joule's work were largely derogatory.[15] Many years later, after he had achieved great fame, Helmholtz offered an apology and an explanation in the form of an addendum to a reprint of that paper:

To the history of the discovery of the law of the conservation of energy [Kraft] it should be added that in 1842 R. Mayer published his paper on "the forces of non-living nature" and in 1845 the one on "organic motion in its connection with metabolism" in Heilbronn. I first learned about them later, and since then never failed to speak of them first in public [reference to lectures in 1854, 1862, 1869], and to defend them to whatever extent possible against Mr. Joule's friends, who were inclined to reject them completely. . . .Very recently, the adherents of metaphysical speculation have attempted to put the stamp of an a priori principle on the law of the conservation of energy, and therefore hail R. Mayer as a hero in the field of pure thought. What they regard as the pinnacle of Mayer's contributions, namely the metaphysically formulated pseudo-evidence for the a priori necessity of this law, will seem to every investigator accustomed to strict scientific method to be just the weakest aspect of his explanations, and is unquestionably the reason why Mayer's work remained unknown so long in scientific circles. Only after the road had been broken open to the conviction that the law is valid from another direction, namely through Mr. Joule's masterly researches, was attention given to Mayer's writings.[16]

To this, Helmholtz added that if his "knowledge of the literature was incomplete in 1847," it was because he was then [on military duty as an army surgeon] in Potsdam, where he had access only to the library of the local Gymnasium.

In his youth, Helmholtz acquired a strong interest in physics and German philosophy, but his father, a teacher at the Potsdam Gymnasium, could not provide the fees for the University of Berlin, and urged his son (the eldest of four children) to apply for a government stipend to study medicine at the local medical-surgical institute. He was awarded the grant, which entailed a commitment to several years of military service after the completion of his medical training. While a student at the institute (1838-1842), Helmholtz continued his self-education in mathematics and philosophy, and also attended the lectures of several professors at the university, including those of Johannes Müller, the professor of anatomy and physiology to whom he was especially attracted. After receiving his M.D. degree, and serving a year's internship at the Charité Hospital, Helmholtz began his military service as an army surgeon in Potsdam in 1843. During 1845-1846, while on leave to prepare for the state medical examinations, he worked in the laboratory of Gustav Magnus on the chemical changes in muscular contraction. In 1848, Helmholtz was released from military duty, and served for several months as Müller's assistant before being appointed associate professor of anatomy and physiology at Königsberg; he was promoted to full professor in 1851. There, he conducted brilliant experiments in physiological optics and invented the ophthalmoscope. In 1855, Helmholtz moved to Bonn, and three years later to Heidelberg, where he made important contributions to physiological acoustics; he also entered the fields of hydrodynamic and electrodynamic theory. During the mid-1850s, he began to present popular scientific lectures, for which he was widely acclaimed. In 1871, he was made professor of physics at Berlin, and returned to thermodynamics in the early 1880s. He was elevated by the Kaiser to the nobility in 1883, and in 1888 was appointed president of the newly established *Physikalisch-Technische Reichsanstalt*.[17]

The transformation during the 1840s of Helmholtz's style of thinking from that of an experimental physiologist to that of a theoretical physicist attests to the catholicity of his scientific interests, and provides a foretaste of his attitude to the later development of the chemical sciences. The 1845 paper on the chemical changes on muscular contraction began:

> One of the most important physiological questions, intimately related to the nature of the vital force, namely whether the life of organic bodies is the effect of a self-generated purposeful force, or the result of forces also active in inanimate matter but specifically modified through the nature of their cooperation, has recently been given a much more concrete form in Liebig's effort to derive physiological conclusions from known physical and chemical laws, namely whether or not mechanical force and the heat produced

in the organism can be derived completely from metabolic changes [Stoffwechsel].[18]

In this paper, Helmholtz reported the results of replicate experiments in which excised frog muscle was subjected to electrical stimulation, and portions of the macerated tissue were then extracted with water, alcohol-water ("spiritus"), or alcohol. The extracts were then evaporated to dryness, and the residues weighed. The results were compared with those obtained after extraction of equal amounts of non-stimulated muscle. He found that "the aqueous extract is diminished in the electrified portion of muscle and, inversely, the spiritus and alcohol extracts are increased in comparison to the non-electrified portion." No attempt appears to have been made to isolate any individual chemical substance, for example lactic acid. Helmholtz also noted that "I must leave unresolved another of the most important problems; whether the muscle fiber participates in the decomposition. A priori this would be quite probable, because we find the protein compounds everywhere as carriers of the highest life energies, and especially in our case the appearance of a sizable amount of sulfate and phosphate salts in the urine after muscular exertion argues for a decomposition of sulfur and phosphorus compounds."[19] The theme of "energy-rich proteins" was reiterated by numerous physiologists and some chemists (for example, Thomas Graham) throughout the rest of the nineteenth century.[20] As for Helmholtz's attitude to the development of organic chemistry during the second half of the nineteenth century, he wrote in a letter in 1891 that "the whole immensely comprehensive system of organic chemistry has developed in the most irrational manner, always with attached sensory pictures which cannot possibly be correct in the way they are presented."[21]

In his *Erhaltung der Kraft*, Helmholtz's style was to affirm a principle, formulate it in mathematical terms, and discuss the applicability of the principle to mechanical, electrical, chemical, and physiological work. He stated the principle as follows:

> In all cases of motion of free material points under the influence of their attractive or repulsive forces, whose intensities depend only on their distance, the loss of the quantity of tension force [Spannkraft] is always equal to the gain of living force [lebendige Kraft], and the gain of the first to the loss of the second. It is therefore the sum of the available living and tension forces which is always constant. We can denote our law in this most general form as the principle of the conservation of force.[22]

Most of the remainder of the paper dealt with the applications of the principle, and was notably deficient in Helmholtz's assessment of Joule's experimental work on the electrical and mechanical equivalents of heat. Helmholtz also discussed the "known natural processes in organic beings" in a manner reminiscent of the speculations of Dumas and Liebig:

> In plants the processes are principally chemical, and at least in some, there is a slight generation of heat. In the main, there is deposited in them an enormous quantity of chemical tension force, whose equivalent is yielded to us as heat upon combustion of plant substances. According to our present knowledge, the only living force absorbed during the growth of plants are the chemical rays of sunlight. . . . [Animals] take up the complex oxidizable compounds produced by plants, as well as oxygen, mostly burned to carbon dioxide and water, partly reduced to simpler compounds, thus using up a certain amount of chemical tension forces, and generate heat and mechanical forces. Since the latter represent a relatively small amount of work as against the quantity of heat, the question of the conservation of force reduces itself

approximately to whether the combustion and conversion of nutrients generates the same quantity of heat as that released by animals. According to the experiments of Dulong and Despretz, this question can be answered, at least approximately, in the affirmative sense.[23]

After the work of Lavoisier and Laplace, especially during 1820-1850, numerous calorimetric measurements of the heat produced in chemical reactions were made. In 1840, Germain Henri Hess concluded that "the heat developed in a chemical change is constant whether the change occurs directly or in several steps."[24] This so-called "law of heat summation" was adopted by Liebig and some of the physiologists who were influenced by his ideas, and the high point of that development came during the 1890s in the work of Max Rubner, who reported that the heat produced by a dog in a respiration calorimeter equaled the heats of the combustion of dog fat and protein, minus that of the organic constituents of dog urine. This "intake-output" approach became popular in the United States during the period 1900-1930, and was termed "energy metabolism."[25]

Thermochemistry also developed in a different direction after 1850. The Danish chemist, Julius Thomsen, who performed many calorimetric measurements of the heat production in a variety of chemical reactions, wrote: "By considering the amount of heat evolved in the formation of a compound as a measure of affinity, as a measure of the work required to cleave the compound into its constituents, it must be possible to deduce general laws for the chemical processes and to replace the old doctrine of affinity, which rests on an uncertain groundwork, by a new one resting on a sure numerical basis."[26]

Marcelin Berthelot, who entered the field after Thomsen, went further, offering what he called the "principle of maximum work," which stated that "every chemical change accomplished without the intervention of external energy tends towards the production of a body or a system of bodies which produces the most heat."[27] Like Thomsen, Berthelot claimed that thermochemical data provided a measure of chemical affinity. There ensued a bitter rivalry, lasting from 1872 until 1886, in which each man accused the other of publishing inaccurate data, and which was exacerbated by Thomsen's irascible temperament. Although both later realized that some spontaneous chemical reactions proceed with the *absorption* of heat, and acknowledged that the principle was strictly valid only at absolute zero, Berthelot was more persistent in affirming the utility and approximate validity of the doctrine.

I return to the development of chemical thermodynamics and the contributions of Rudolf Clausius. Three years after he received his Ph.D. in mathematics and physics at Halle, he published his famous paper bearing in its title the German translation of the title of Carnot's memoir.[28] Clausius's first teaching appointment was at the Royal Artillery and Engineering School in Berlin. In 1855, he became professor of mathematical physics at the new technical university in Zürich, in 1867 he moved to Würzburg, and in 1869 to Bonn, where he remained the rest of his life.[29]

In his 1850 paper, Clausius presented a mathematical analysis of the Carnot cycle, and took account of Joule's experimental results in correcting Carnot's assumption that no heat (in the form of caloric) is lost in the process. By considering heat to be a form of motion, not a material substance, Clausius concluded "that in all cases where work arises from heat, a quantity of heat proportional to the generated work is used up, and conversely through the use of an equivalent amount of work, the same quantity of heat can be generated."[30] In this paper, Clausius referred to the "painstaking" [sorgfältige] experiments of Joule and Regnault, and the writings of Mayer and Thomson, but did not mention Helmholtz's name, although some of Helmholtz's intuitive assumptions appear in more precise mathematical form. In subsequent articles, Clausius developed what came to be known as the second law of thermodynamics. In one, published in 1865, he replaced what he had called "equivalent value" with the term "entropy," which he later defined as "a magnitude which represents the sum of the transformations which must have taken place in order to bring any body or system of bodies into its present condition,"[31] and offered his famous maxim: "The energy of the universe is constant. The entropy of the universe tends to a maximum." Clausius also introduced the term "disgregation" to denote changes in molecular arrangement. Like entropy, this term received a mixed reception, but unlike entropy, it did not survive into the twentieth century. The last decades of the nineteenth century were marked by confusion and misunderstanding about these terms, with some British physicists (notably Peter Guthrie Tait) strongly critical, and the debate assumed a nationalistic tone. Some German physicists rose to the defense of Mayer's claims for priority in the formulation of the first law, and disputed British claims for Thomson's priority in the statement of the second law. It was a time when opinion was divided regarding the nature of the motion of the elementary particles constituting matter, with some physicists favoring Rankine's idea of the rotating vortex model of the atom, and others favoring Clausius's idea of periodic collisions of atoms moving in straight lines.[32]

During the course of his elaboration of the concept of entropy, Clausius made contributions to the development of the kinetic theory of gases. In his 1850 paper, like Joule and Helmholtz before him, Clausius adopted the idea of heat as a mode of motion and a form of energy that can be converted to mechanical or electrical work. Earlier in the nineteenth century, the kinetic theory (suggested by David Bernouilli in 1738) had been treated by John Herapath during 1816-1821, and by John James Waterston in 1846.[33] Joule took account of Herapath's contribution, but Waterston's was not appreciated until the 1890s. Clausius added to the theory the idea of the mean free path, namely the average distance traveled by a molecule between successive collisions; in 1860 James Clerk Maxwell refined this idea by introducing the concept of the distribution of velocities.[34] In his *Theory of Heat* (1871), he derived his thermodynamic theory by purely geometrical reasoning.

After his studies at Cambridge, Maxwell began his academic career in 1856 at Marichal College, Aberdeen, then moved to King's College, London (1860-1865); in 1871 he was appointed the first Cavendish professor at Cambridge. An outstanding experimenter, Maxwell became famous for his insight in the elaboration of thermodynamic theory ("Maxwell's demon"), his studies on electricity and magnetism, and the electromagnetic theory of light (of which more later). To this must be added his imaginative use of mechanical models and the lucidity of his writings.[35]

The capstone of the kinetic theory of gases was provided by Ludwig Boltzmann, a pupil of Joseph Loschmidt and, at various times, professor of physics at Vienna, Graz, Munich, and Leipzig. Boltzmann's most famous contribution was the extension of Maxwell's theory of the distribution of energy, and the demonstration in 1877 that entropy is related to the statistical distribution of molecular configurations and that an increase in entropy corresponds to increased molecular randomness. Some years later, Boltzmann's approach was refined by Josiah Willard Gibbs (of whom more shortly), who was also an admirer of Clausius, and who gave a new term, *Statistical Mechanics*, in the title of his book (1902) on this subject. Another admirer of Boltzmann's work was

Max Planck, whose concept of the "partition function" was derived from Boltzmann's distribution law. Other German and Austrian colleagues, some of whom objected to his atomistic leanings, gave Boltzmann little support. He committed suicide in 1906 for reasons that are not clear, but his despair at this lack of recognition may have been a contributing factor.[36]

In what has gone before, I have attempted to sketch the work and style of several noted German and British nineteenth-century physicists who participated in the development of thermodynamic theory, and the formulation of the concept of entropy. It must be emphasized, however, that this development did not have an appreciable impact on the work of experimental chemists until the 1880s. One of the notable exceptions was August Friedrich Horstmann, to whose work brief reference is regularly made, but who merits closer historical study. He was a member of a well-to-do Mannheim family, received his Ph.D. in 1865 at Heidelberg for work in Bunsen's laboratory, and continued his studies in Zürich, where he attended the lectures of Clausius and Wislicenus. Horstmann then became a Privatdozent in Heidelberg. Afflicted from his youth by progressive eye disease, he was unable to embark on a career leading to a professorship, but his financial situation allowed him to do intensive research in his private laboratory until about 1889, when he became totally blind. The persistence with which Horstmann coped with his disability, which caused him to drop glassware and spill mercury on the floor, is all the more admirable in light of the importance of his contribution to the development of chemical thermodynamics. He lived for another 40 years, comforted by the sound of classical music.[37]

In a series of papers published between 1869 and 1881, Horstmann reported the results of his studies on the effect of temperature on the equilibrium constant in the dissociation of substances such as ammonium chloride ($NH_4Cl = NH_3 + HCl$) or phosphorus pentachloride ($PCl_5 = PCl_3 + Cl_2$). It had been found earlier that the vapor arising from the evaporation of the melted form of these substances was "anomalous" in the sense that the vapor density was much less than what was expected on the basis of the Avogadro hypothesis. In 1857, Sainte-Claire Deville suggested that this behavior was a consequence of the "dissociation" of the substance into smaller particles, and this view received support in the experiments of Leopold von Pebal (1862). Horstmann showed that the dissociation pressure of ammonium chloride (for which melting and vaporization occur at the same temperature) follows the same law as that for the vapor pressure of a liquid, and that the heat of dissociation may be substituted for the heat of evaporation in the Carnot-Clapeyron equation. He also formulated an equation for the dissociation of PCl_5, based on Clausius's principle that at equilibrium, when the evaporation of a liquid is exactly equal to the reabsorption of the vapor, the entropy is at a maximum. Horstmann also suggested that "in the case of dilute solutions, the disgregation of the salt depends in a similar way on the distances of the molecules as in a permanent gas, an assumption that has some probability."[38]

In the preface to his *Études de Chimie Dynamique* (1884), van't Hoff acknowledged the influence of Horstmann's writings on his own studies of chemical equilibrium and the theory of solutions, and in 1885 Ostwald wrote in his textbook that "if the idea-rich papers of Horstmann have not brought the appropriate fruits, there is no doubt that they will lead to the future elaboration of a theory of affinity based on purely mechanical principles." In the second edition (1902) he added: "The progress of science has confirmed this prediction more rapidly than could have then be expected" and wrote that "[this] foundation of the study of affinity on the second law of thermodynamics is of the greatest significance and, after having received little attention, has undergone a great and brilliant development.[39] On the other hand, in 1883, Helmholtz credited Lord Rayleigh with the first application of the entropy principle to chemical processes; in 1875, Rayleigh stated that "the chemical bearings of the theory of dissipation . . . have not hitherto received much attention."[40]

If Horstmann's experimental efforts were only partially effective in bridging the intellectual gap that separated physicists from chemists, it was Josiah Willard Gibbs, a Connecticut Yankee and an isolated eminence in a scientific prairie, who built during the 1870s a mathematically grounded bridge between the two kinds of scientists. His papers, *On the Equilibrium of Heterogeneous Substances* (1876-1878), totaling 321 pages, were published in the local (though widely distributed) *Transactions of the Connecticut Academy of Sciences*. Although these papers were known in Germany, Britain, and France after 1880, and admired by Maxwell, Ostwald, and Le Châtelier, Gibbs's writings were difficult for chemists to understand, in large part because of his abstract, strictly logical, style of presentation, and because he made only scant reference to experimental data. It required the insight and scientific work of men such as van't Hoff (and much later, Gilbert Newton Lewis) to make Gibbs's ideas and method of reasoning accessible to working chemists.[41]

Of particular historical interest is Pierre Duhem's appreciation of Gibbs on the occasion of the publication of Gibbs's collected papers in 1906.[42] Duhem had been introduced to chemical thermodynamics by his teacher, Jules Moutier, who was working during the 1880s in the laboratory of Henri Sainte-Claire Deville at the Ecole Normale Supérieure. In 1908, Duhem was a professor at Bordeaux. Like Laurent, who had incurred the displeasure of Dumas, Duhem was not given a professorship in Paris, because of his opposition to Berthelot's principle of maximum work.[43]

One of five children, Gibbs was born in New Haven, the only son of a noted Yale professor of philology and sacred literature. After graduating from Yale College in 1858 and receiving his Yale Ph.D. in engineering in 1861, young Gibbs taught Latin and physics to Yale students. He then spent three years (1866-1869) in Paris, Berlin, and Heidelberg, where he attended the lectures of famous mathematicians and physicists, among them Helmholtz and Kirchhoff. Two years after his return to New Haven, Gibbs was appointed professor of mathematical physics at Yale, but he did not receive a regular salary until 1880, when he was offered a professorship at the newly established Johns Hopkins University. Gibbs remained at Yale the rest of his life, did not travel abroad, never married, and lived out his years in the family home. A friendly but private man with serene self-confidence, Gibbs pursued his studies with single-minded determination, always seeking greater simplicity and generality in the formulation of scientific principles. Unlike most of his European counterparts, he never engaged in public squabbles over matters of priority, but at least on one occasion he felt obliged to respond in the pages of *Nature* to an attack on his use of vector analysis in preference to quaternions.[44]

Gibbs's first published papers (1873) dealt with the improvement of geometrical representation of thermodynamic principles, and the formulation of what he called "the fundamental equation of the fluid." Starting with Clapeyron's graphic method, and Clausius's formulation of the second law, Gibbs arrived at his famous equation $d\varepsilon = td\mu - pdv$ (where ε is the internal energy, μ the entropy, t the absolute temperature, p the pressure, and v the volume). In deriving this equation, he defined $td\mu$ as the heat (dH) and pdv as the work (dW) in passing from one state of a body to another. The equation is now usually written as $\Delta G = \Delta H - T\Delta S$, where G is the Gibbs's free energy, H the enthalpy, and S the entropy. In a second paper, published in the same year, Gibbs enlarged the scope of geometrical representation of thermodynamic equilibrium by using surfaces to define the states of such equilibria in multiphase systems.[45]

In his two-part *On the Equilibrium of Heterogeneous Substances*, Gibbs showed how his approach could be applied to a variety of chemical, surface, and electrochemical phenomena. For example, he used the concept of "chemical potential" to derive the "phase rule," which assumed great importance when it was adopted as a basis for an extensive experimental program by a group of Dutch chemists led by Bakhuis Roozeboom.[46] Gibbs did not mention the work of Horstmann, but did refer to François Massieu, who had proposed characteristic thermodynamic functions in 1869.

In his final years, Gibbs developed an approach to statistical mechanics that was found by many mathematical physicists to be more useful than that of Boltzmann.[47] It should be added that not all of Gibbs's writings were published during his lifetime. Indeed, his very first papers, dealing with such practical engineering problems as the design of an improved railroad car brake (for which he received a patent in 1866) and of governors for steam engines, were not published until 1947.[48]

Hermann Helmholtz re-entered the field of thermodynamics in 1882, after he had abandoned his adherence to the Thomsen-Berthelot principle that the tendency of a chemical reaction to occur spontaneously depends on the heat evolved in the reaction, and it had been shown that some endothermic reaction can occur spontaneously. Using Clausius's treatment, he derived an equation, which may be written in a more modern form as $\Delta A = \Delta U - T\Delta S$, where A is the "free energy," U is the total energy, and $T\Delta S$ is the "bound energy."[49] According to this equation, at constant temperature and volume, a reaction can occur spontaneously only when there is a decrease in A. This function is different from Gibbs's free energy, which specifies the condition for spontaneity at constant pressure. For a time, the two equations were lumped together as the "Gibbs-Helmholtz" equation.

In this paper, Helmholtz did not refer to Gibbs, who is reported to have sent him reprints of the papers published during the 1870s, nor to Horstmann. Helmholtz partly corrected his error in 1883, when he referred to Gibbs and Massieu, but he never mentioned the work of Horstmann. He also claimed that his electrical studies during the 1870s led him to develop his equation.[50]

The mode of van't Hoff's entry into chemical thermodynamics was different from that of Gibbs or Helmholtz. After his stereochemical studies, van't Hoff initiated in his Amsterdam laboratory an intensive program of experimental work on the equilibria and kinetics of chemical reactions.[51] He outlined the program in a two-volume book published in 1877 and 1881,[52] and summarized the initial results in another book published in 1884.[53] The latter book was translated into German and English soon afterward, and was chiefly responsible for the wider recognition by working chemists of the importance of thermodynamic principles. The mathematical treatment was less demanding or rigorous than that of Gibbs's, and greater attention was given to experimental data. An enlarged three-volume version, which took account of later work, was published in 1898-1900.[54]

In 1877, van't Hoff corrected the mathematical formulation by Berthelot and Péan de Saint-Gilles of the equilibrium in the esterification of acetic acid by ethyl alcohol

($CH_3COOH + C_2H_5OH \rightarrow CH_3CO\text{-}OC_2H_5 + H_2O$), by considering the equilibrium to be the result of the velocities of both the forward reaction and the reverse hydrolytic reaction. Van't Hoff defined the equilibrium constant for the reaction $A + B \rightleftharpoons C + D$ as $K = [C] [D]/[A][B] = k_1/k_{-1}$ where the brackets denote the molar concentrations and k_1 and k_{-1} are the rate constants for the forward and reverse reactions.[55] This kinetic approach, which did not invoke the concept of entropy, was an outgrowth of the polarimetric studies of Ludwig Wilhelmy on the rate of the inversion of sucrose by acids, of the work of Vernon Harcourt and William Esson on the rates of such reactions as that between hydrogen peroxide and hydrogen iodide, and of the so-called "law of mass action" proposed by Cato Maximilian Guldberg and Peter Waage.[56]

Van't Hoff extended and generalized the mathematical analysis of chemical kinetics in terms of the number of molecules entering into a reaction, and introduced a method for determining its "order" (monomolecular, bimolecular, etc.). He then introduced an important equation for the temperature dependence of an equilibrium constant (in modern notation) as $d\ln K/dT = \Delta\ln U/RT^2$. He also laid the groundwork for the so-called Arrhenius equation ($\ln k = Ae^{-E/RT}$) for the temperature dependence of the rate constant. Moreover, he showed that for the conversion of $A + B$ into $C + D$, the "work of affinity" is given by the equation $w = RT\ln K - RT\ln([C][D]/[A][B])$. He also expressed the work of affinity as $w = q[(T_t - T)/Tt]$, where q is the heat evolved and T_t is the temperature at which $K = 1$, and thereby derived an equation equivalent to the so-called Gibbs-Helmholtz equation without involving entropy.

To these accomplishments van't Hoff added the finding that the data obtained by the botanist Wilhelm Pfeffer for the pressure on a semipermeable membrane by a substance such as sucrose dissolved in water accorded with the equation $\pi V = nRT$, where π is the pressure, n is the molar amount of solute (n/V = molar concentration), a relationship analogous to that of an ideal gas ($PV = nRT$).[57] Osmotic pressure is one of the "colligative" properties, the others being the depression of the freezing point, the elevation of the boiling point, and the lowering of the vapor pressure. The finding that substances that conducted electricity gave abnormally high values of π was explained by means of the theory of electrolytic dissociation, advanced by Svante Arrhenius during the 1880s.

Arrhenius is usually considered to have been a member of the famous triumvirate that established physical chemistry as an independent university discipline. At the start of his scientific work, however, it seemed certain that he would have no university career at all.[58] After his studies at Uppsala (1876-1881) and research in Stockholm (1881-1883), Arrhenius presented at Uppsala a dissertation, published in 1884, which was severely criticized by his two chief professors, Per Theodor Cleve and Otto Petterson. The dissertation dealt with the measurement of the electrical conductivity of dilute solutions of various chemical compounds as an approach to the determination of their molecular weights and reactivity. Ostwald was then studying the chemical affinity of acids by other methods, and immediately switched to conductivity measurements.[59] Arrhenius's oral defense was ineffective, and the dissertation did not receive the *summa cum laude* rating required for an appointment at Uppsala as Dozent. This setback might have obliged Arrhenius to accept the life of a teacher at a *Gymnasium*. Ostwald's interest in the dissertation, however, led him to offer Arrhenius a post in his laboratory in Riga, and this action persuaded the Uppsala authorities to appoint Arrhenius as a Dozent.

In the background of this work was the theory offered in 1851 by Alexander Williamson that, in solution, the molecules of an acid are partly decomposed into smaller units. This idea was later extended by Clausius to apply to electrolytes, and the outstanding experimental work of Wilhelm Hittorf and Friedrich Kohlrausch showed that the velocity and conductivity of ions were distinctive and additive properties.[60]

In his dissertation, Arrhenius reported the empirical observation that the conductivity of a solution of an electrolyte increased upon increased dilution, indicating an increase in the proportion of the dissociated species. After the publication of van't Hoff's *Cours de Dynamique Chimique,* Arrhenius found a guide to the more convincing formulation of his theory, published in 1887.[61] He also was stimulated by van't Hoff's treatment of the temperature dependence of reaction rate constants, and in the equation given above, used the term "energy of activation"[62] (E), a concept of considerable importance in the later development of reaction kinetics.

After his appointment as Dozent, Arrhenius received a travel grant for study with Ostwald, Kohlrausch (in Würzburg), Boltzmann (in Graz) and van't Hoff (in Amsterdam), and then became lecturer (1891) and professor (1895) at the Stockholm technical high school. In 1901, Arrhenius was elected (after strong opposition) to the Swedish Academy of Science, was awarded the Nobel Prize in Chemistry in 1903, and in 1905 he became director of the Nobel Institute for Chemistry, where he remained the rest of his life. He was a member of the Academy committee charged with the selection of recipients of the Nobel Prize in Physics, and was also active as an *éminence grise* in the decisions of the committee for the prize in chemistry.[63]

At the turn of the century, Arrhenius sought to apply physical-chemical ideas to study of cosmic physics, volcanoes, and auroras, but devoted most of his scientific effort to what he named "immunochemistry." In a collaboration with the Danish bacteriologist Thorvald Madsen, he reported experiments that he interpreted as evidence for the view that the neutralization of a toxin by an antitoxin is analogous to the reversible reaction between an acid and base. This view drew the fire of Paul Ehrlich, whose more complex "side-chain" theory was based on the biochemical ideas of the time, in particular the "lock and key" hypothesis of Emil Fischer.[64]

If, among the three so-called founders of physical chemistry, van't Hoff stands forth as the principal source of inspiration for the theoretical and experimental advances that followed the publication of his *Études de Dynamique Chimique*, it was Wilhelm Ostwald who, through the multiple variety of his personal researches and his seemingly inexhaustible entrepreneurship, brought about the acceptance of physical chemistry as an independent academic discipline. A Baltic German, Ostwald studied chemistry with Carl Schmidt (a pupil of Liebig and Wöhler) at Dorpat (now Tartu), where he received his doctoral degree in 1878. Three years later, he became professor of chemistry at the Riga Polytechnic Institute, and in 1887 was appointed to the first German professorship of physical chemistry in Leipzig; he retired in 1906 in order to pursue his other interests. In 1887, he founded the *Zeitschrift für physikalische Chemie*, with van't Hoff and Arrhenius as co-editors. Ostwald also displayed his capacities as a teacher in his *Lehrbuch der allgemeinen Chemie* (1885-1887) and other textbooks. His keen interest in the history of science was evident in his *Elektrochemie: Ihre Geschichte und ihre Lehre* (1903), and his initiation of the series known as *Ostwald's Klassiker*. From his youth, Ostwald was devoted to literature and music (he played the piano and the viola), and he enjoyed painting. In later life, he advocated pacificism and internationalism,[65] but was one of the signers of the notorious *Aufruf an die Kulturwelt* of 4 October 1914, disclaiming German responsibility for the outbreak of the war, defending the invasion of Belgium, and denying the reports of atrocities there.[66]

In Dorpat, Ostwald developed an interest in the use of physical methods to define chemical affinity in quantitative terms. His first efforts involved the dilatometric measurement of volume changes in the neutralization of acids by bases. He then turned to a study of the effectiveness of acids as catalysts of various reactions, including the hydrolytic cleavage ("inversion") of sucrose.[67] After the appearance of Arrhenius' report in 1884, and of van't Hoff's *Études de Dynamique Chimique*, Ostwald used electrical conductivity as the method of choice, found that it increases upon dilution of aqueous solutions of monobasic acids, and proposed what he called the "dilution law," which connects the conductivity to the equilibrium constant for the dissociation of an acid into a pair of ions.[68]

Ostwald's research group in Leipzig carried forward the main lines of his experimental work during the 1880s, and made valuable contributions to the development of electrochemistry. For example, in 1898 his student, Rudolf Peters, provided the experimental basis for the comparative ability of electromotive substances to act as oxidants or reductants.[69] Also, in a discussion of oxidation reactions, Ostwald made a distinction between the transfer of chemical energy and the transfer of heat or electrical energy:

> A direct transformation of chemical energies is only possible to the extent that chemical energies can be set in connection with each other, that is, within such processes which are represented by a stoichiometric equation. Coupled reactions of this kind may be distinguished from those which proceed independently of each other; their characteristic lies in the

fact that they can be represented by a single chemical equation with definite integral coefficients.[70]

The concept of coupled oxidation-reduction reactions was of considerable importance in the development of biochemical thought after 1920. On the other hand, Ostwald's attempt to develop a fruitful theory of catalysis was unsuccessful.[71]

During the 1890s, Ostwald's Leipzig institute attracted many students, of whom 24 were from the United States (including W.D. Bancroft, A.A. Noyes, and T.W. Richards) and eight from Great Britain (including F.G. Donnan). Although the Americans played a large role in the institutional development of physical chemistry in their home country,[72] the Englishmen, led by Henry Armstrong, encountered strong opposition to the theory of solutions propagated from Leipzig.[73] Despite the productivity of his junior associates, Ostwald appeared to lose much of his earlier enthusiasm for experimental work, and began to advocate "energetics" as a substitute for materialism and atomism. As he put it in 1904:

> It is possible to deduce from the principles of chemical dynamics all the stoichiometrical laws; the law of multiple proportions and the law of combining weights. You all know that up to the present time it has only been possible to deduce these laws by the help of the atomic hypothesis. Chemical dynamics has, therefore, made the atomic hypothesis unnecessary for this purpose and has put the theory of the stoichiometric laws on more secure ground than that furnished by a mere hypothesis.[74]

In 1909, after Jean Perrin's demonstration of the reality of atoms, Ostwald wrote that "the atomic hypothesis is thus raised to the position of a scientifically well-founded theory," but added that "from the point of view of stoichiometry the atomic theory is merely a convenient mode of representation, for the facts, as is well known, can equally well, and perhaps better, be represented without the aid of the atomic conception as usually advanced."[75]

Ostwald resigned his professorship in 1906, and spent the remaining 26 years of his life at his estate in Grossbothen, where he developed a valuable system of color standards and harmony, and wrote extensively on various topics, including monism, the energetic imperative, the pyramid of the sciences, and happiness.[76] He also composed a three-volume autobiography, with comments about some of his contemporaries. For example, he wrote about an encounter at an 1889 meeting:

> I found myself in a swarm of organic chemists gathered about Emil Fischer, already regarded as the future leader of our science, since what was not organic chemistry was not recognized as chemistry. To his disparaging remark about our new direction I answered that the organic chemists owe us thanks for the possibility of determining the molecular weights of non-volatile substances. Fischer replied: "This was entirely unnecessary; I see directly the molecular weight of every new substance, and do not need your methods.". . . I have since then had repeated evidence of that unconditionally negative attitude.[77]

Fischer's "negative attitude" toward Ostwald, and to the use of physical-chemical methods in his research, was well known, but Fischer also held van't Hoff and Walther Nernst in high regard, and helped to bring them to Berlin.

Like van't Hoff before him, Nernst sought to apply the physics of his time to the investigation of fundamental chemical problems. The difference between them was that van't Hoff had begun as a brilliant organic chemist, and that Nernst began as a brilliant experimental physicist. After graduating from *Gymnasium* at the top of his class, Nernst studied physics during 1883-1887 at Zürich (Weber), Berlin (Helmholtz), Graz (Boltzmann, Ettinghausen), and obtained his doctorate in the latter year at Würzburg (Kohlrausch) for work begun in Graz with Ettinghausen. At Würzburg he came to know Emil Fischer and, of more immediate importance, Arrhenius, who introduced Nernst to Ostwald and the research program of the so-called "ionists." There followed an assistantship in Leipzig, and Nernst's *Habilitation* in 1889. In 1891, he became associate professor at Göttingen, was promoted to full professor of physical chemistry in 1894, and in 1905 moved to Berlin to head the physical-chemical institute there. Except for a short service (1922-1924) as president of the *Physikalisch-technische Reichsanstalt*, he remained at the university until his retirement in 1934.[78]

Nernst demonstrated his promise as an investigator in his doctoral dissertation, a study of the electromotive forces elicited by magnetism in heated metallic electrodes, and then examined the connection between diffusion and conductivity by measuring the the effect of potential gradients on the rate of ionic motion. In his *Habilitationschrift*, which won him great fame, Nernst applied thermodynamic reasoning to his experimental data on the connection between electromotive force and ionic concentration.[79] He showed that, in dilute solution, the electromotive force E for a galvanic process involving a change in concentration from C_1 to C_2, $E = (RT/N)\mathscr{F}\ln(C_1/C_2)$, where R is the gas constant, T the absolute temperature, N the gram equivalents, and $\mathscr{F}$ the Faraday constant. Nernst then turned his attention to the distribution of solutes between two immiscible fluids.[80]

In 1891, when Nernst was being considered for promotion to associate professor at Göttingen, Friedrich Althoff, the czar of Prussian higher education, asked Helmholtz for his opinion. In his prompt reply Helmholtz wrote:

I know Dr. Nernst partly from his papers, partly from his several visits to me. I have the impression that he is a very serious, sharp-witted, and thoughtful worker. He has thrown himself wholeheartedly into the newest trends of physical chemistry which originated with the Hollander van't Hoff and defended with great enthusiasm in his journal by Professor Ostwald in Leipzig, with whom Nernst is an assistant. These theories have already shown themselves in manifold ways to be highly fruitful, and have led to really correct conclusions, although they contain arbitrary assumptions which seem to me not to have been proved. However, the chemists need these assumptions (on the dissociation of a part of compound molecules of dissolved salts) in order to build a visual representation of the processes. . . . There is in this entire trend a healthy nucleus, the application of thermodynamics to chemistry, which Planck has also put forward in a more exact form. But the thermodynamic principles in their abstract form are only to be grasped by rigidly trained mathematicians and are therefore difficult to comprehend by people who must perform experiments on solutions, and their vapor pressures, freezing points, heats of solution, etc. There is therefore needed an intermediary group of teachers who know how to present the abstract principles in a more concrete form, and among them Nernst is, in any case, one of the ablest.[81]

As was suggested earlier, Helmholtz, as a physiologist turned physicist, took a dim view of the development of chemistry after 1850.

In his Göttingen institute, Nernst's group was engaged in electrochemical research on storage batteries, polarization, overvoltage, electrode potentials, and electrocapillarity. He developed a new method for the measurement of the dielectric constant of fluids by means of an alternating current Wheatstone bridge. Instruments were devised and constructed to study reactions at very high temperatures and pressures. Improved apparatus of various kinds—microbalances, calorimeters, transformers—were also fabricated at the institute.

These technical achievements, some of considerable industrial importance, attest to Nernst's ingenuity and skill in instrumentation. His studies on the conductivity of solid substances at high temperatures led to the design of a lamp in which the carbon filament was replaced by one composed largely of zirconium oxide; Nernst derived a considerable income from the sale of the patents, although his invention was supplanted soon afterward by the tungsten lamp.

During his final years at Göttingen, Nernst also formulated what was to be the high point in his contributions to thermodynamic theory—his "heat theorem" (he lovingly called it "mein Wärmesatz"), also termed the third law of thermodynamics.[82] As is indicated in the title of the paper, Nernst formulated this theory as a means of calculating the equilibrium constant (K), and hence the free energy change (ΔF), from calorimetric data for the change in heat content (ΔH). According to the Thomsen-Berthelot principle, based on the first law, $\Delta F = \Delta H$, and after 1850 a large body of calorimetric data had been accumulated for values of ΔH. It was found that in those cases where the value of ΔF was also known (from electrometric measurements or from the determination of equilibrium concentrations), ΔH usually did not equal ΔF, although in some cases the two values were nearly the same. After the formulation of the second law, and its representation in the form of the so-called Gibbs-Helmholtz equation, several investigators, notably Henri Louis Le Châtelier, Gilbert Newton Lewis, Theodore William Richards, and Fritz Haber,[83] attempted to solve the problem posed by the impossibility of obtaining, by calorimetric measurement, the value of the constant in the integration of the term for $-\Delta S = (\delta \Delta F / \delta T)$. Nernst's proposed solution of the problem was to assume that, for solids and fluids, the curves for the temperature dependence of free energy and heat content come together asymptotically as the temperature approaches absolute zero, and his heat theorem states that $\Delta F = \Delta H_o - \beta T^2 - (\gamma/2)T^3$ - etc.; since the coefficients β, γ, etc. can be calculated from the specific heats of the reactants and products at temperature T, the value of ΔF can be calculated from calorimetric data. After 1906, the main experimental effort in Nernst's new Berlin institute was directed to the accumulation of data to test the validity of the heat theorem, in particular the determination of specific heats and their temperature coefficients at very low temperatures. This work, initially performed by a large group which included Frederick Lindemann (later Lord Cherwell), and after 1920 Franz (later Sir Francis) Simon, involved the use of specially designed new instruments for the electrical measurement of temperature changes, for vacuum calorimetry, and for the liquefaction of hydrogen. After World War II, the low temperature studies were continued at Oxford, and temperatures as low as 10^{-5} degrees Kelvin were attained. As the evidence in support of the theorem increased, so did Nernst's confidence in claiming that it represented the third law of thermodynamics. During the years before his retirement, Nernst also made contributions in photochemistry and cosmology.

After 1906, Nernst recognized the relevance of his heat theorem to Albert Einstein's 1905 paper on specific heats (published along with his more famous papers on special relativity and radiation theory), and to Max Planck's quantum theory.[84] The discussion that followed led Nernst to propose to Ernest Solvay, a wealthy Belgian industrialist, that a conference be convened in 1911 to discuss the contemporary problems of physics raised by these developments.[85]

Although the heat theorem came to be enshrined in textbooks, it was not widely used for its intended purpose because spectroscopic methods were found to be more practical. One of the early applications of the theorem was reported in 1913 by Julius Báron and Michael Polanyi, who calculated the free energy in the combustion of glucose at 37°C, and found it to be about 13 percent higher than the calorimetrically determined heat of combustion.[86]

Nernst was widely admired for his originality and inventiveness, but there were adverse comments about his egocentricity and talkativeness. Although generous and helpful to his scientific colleagues, he could be a severe critic, as in his negative reaction to Arrhenius's application of physical chemistry to immunology.[87] A German patriot, Nernst was among the signers of the 1914 *Aufruf*, and the father of two sons killed during World War I, but he strongly opposed the anti-Semitic attacks on Einstein during the 1920s and the later rise of the Nazis to power. In 1934, he left Berlin and lived out his remaining years at his estate in Zibelle, where he enjoyed being a carp farmer.

Nernst's widely adopted textbook on theoretical chemistry went through 15 editions (including translations); the one (with Schönflies) on mathematical methods in scientific research was also extremely popular.[88]

Chapter Nine

ELECTRONS, REACTION MECHANISMS, AND ORGANIC SYNTHESIS

❧

IN THIS CHAPTER I return to the historical development of electrochemistry, culminating in the discovery of the electron, and the subsequent transformation of the concepts of valency and the chemical bond. Having sketched in a previous section the contributions of Davy and Berzelius, based on Volta's invention of the electric pile, I begin with the work and thought of Michael Faraday.

Few figures in the history of the natural sciences have evoked greater admiration than that accorded Faraday. His numerous biographers have attested to his experimental skill, his extraordinary devotion to the task at hand, whether in the laboratory or in the lecture hall, the significance of his achievements, and the generosity of spirit he displayed in his scientific and social relationships.[1]

Both chemists and physicists have claimed Faraday as their own, although he preferred to identify himself as an "experimental philosopher." His title at the Royal Institution was Fullerian Professor of Chemistry, and his first achievements, during the 1820s, were in what came to be called "organic chemistry." It has been argued, in my opinion rightly, that Faraday's style of work and thought in his subsequent studies on electricity and magnetism was more like that of nineteenth-century chemists than that of the mathematical physicists of his time.[2] Faraday's biographers have also questioned the extent to which his scientific thought was influenced by his religious beliefs. Faraday's parents had been members of a nonconformist group known as the Sandemanians, and he formally joined that church in 1821 when he married Sarah Bernard, whose family also were members. The tenets of the sect included the literal acceptance of the teachings of the English Bible, belief in a lay clergy, the commitment to a simple and upright life, the succor of the poor and the sick, and eschewal of the accumulation of personal wealth. In his later life, Faraday occasionally gave Sunday sermons to the congregation, declined high honors (such as the presidency of the Royal Society), and did not patent his inventions. As for its influence on his scientific thought, it would seem that Faraday's religion, like that of many English natural philosophers since Robert Boyle, buttressed the conviction that the purpose of experimentation was to demonstrate the wisdom of God in the creation of an orderly universe.[3]

Before Faraday became Davy's laboratory assistant in March 1813, his education was largely derived from reading (and note-taking) of books that he bound over a period of seven years as an apprentice of a London bookseller. No doubt the manual skill young Faraday acquired as a bookbinder served him well in his later laboratory work. In 1808, he also joined a group that met weekly to discuss scientific subjects. The decisive educational experience, however, came later in 1813, when Davy resigned as head of the Royal Institution (he was replaced by Thomas William Brande), and invited Faraday to accompany him on an extended tour of the continent. The party left England in October 1813 and returned in April 1815, a period during which the war of the allies against Napoleon was still under way, but Davy's fame opened normally closed national borders. The trip included visits (as well as experimental work) to Paris and other French cities, Italy (Turin, Genoa, Florence, Rome, Venice), Germany (Munich), Switzerland (Zürich, Berne, Geneva), and Belgium (Brussels). The daily association with Davy, and the opportunity to meet some of the leading European scientists, provided Faraday with an educational experience superior to any he might have received at a major British university. As was his habit, he took meticulous notes, and, from his accounts in his letters home, it appears that, for Faraday, the only unpleasant feature was Lady Davy's haughty treatment of him.[4]

Upon his return to London, Faraday was prepared to resume his career as a bookbinder, but at Davy's urging he was appointed assistant to Brande at the Royal Institution, where he was given modest living quarters that he did not leave until his retirement in 1862. The years after his reappointment were busy ones. In 1816, he helped Davy with the invention of the miner's safety lamp, and by 1820 he had acquired a reputation as an excellent analytical chemist. Although Davy continued to help Faraday in the composition of his scientific papers, their relationship cooled markedly after 1820, because of misunderstandings about Faraday's failure to make explicit Davy's role in his work on the liquefaction of gases, and to cite Wollaston's earlier work on electromagnetic rotation.[5] In 1824, Faraday was elected Fellow of the Royal Society over the objection of Davy, then president of that body. In the following year, Faraday became Director of the Laboratory, and in 1834 he was appointed Fullerian Professor, making him the second professor at the Royal Institution.

Except for his papers during 1821-1823 on electromagnetic rotation and on the liquefaction of gases, most of Faraday's published writings before 1831 dealt with a great variety of chemical topics. A few of these chemical papers are of exceptional interest in showing his skill as a chemical operator and analyst. He demonstrated that the reaction of "olefiant gas" (ethylene, C_2H_4) with chlorine in sunlight yields a "perchloride of carbon" (C_2Cl_6), which on heating, is converted to a "protochloride of carbon" ($C_2H_4Cl_2$). As was noted in an earlier chapter, such chlorination of hydrocarbons was a prominent feature of the work of Dumas and Laurent during the 1830s on substitution reactions. Faraday also isolated isobutylene (C_4H_8), and called attention to the fact that in this compound the ratio of carbon to hydrogen was the same as that in ethylene; Berzelius later cited this finding as an example of "polymorphism." Another compound isolated by Faraday from oil-rich gas was "bicarburet of hydrogen" (benzene, C_6H_6); he also found what is now called isoprene (C_5H_8) as a product of the distillation of rubber.[6] His analysis of naphthalene, isolated from coal tar by Brande, showed it to have the composition $C_{20}H_8$ ($C = 6$). In 1827, Faraday published his book *Chemical Manipulation*, a practical manual that provides an insight into the methods and apparatus he used in his chemical research.[7] At Davy's behest, Faraday was obliged to do extensive research on the improvement of optical glass before 1831. Later, there were occasional brief papers on such subjects as Schönbein's ozone and on Brownian movement.

Faraday's most famous researches, on electromagnetism and electrochemistry, began in 1822 with the discovery of electromagnetic rotation[8] (Fig. 14). Before his reentry into this field in 1831, there had been considerable work on electromagnetism by François Arago, André Marie Ampère, and Jean Baptiste Biot in France, by John Herschel and William Sturgeon in Britain, and by Georg Ohm and Johann Schweigger in Germany. In the United States, Joseph Henry had discovered electromagnetic induction slightly before Faraday, but published it later, and it was claimed that Fran-

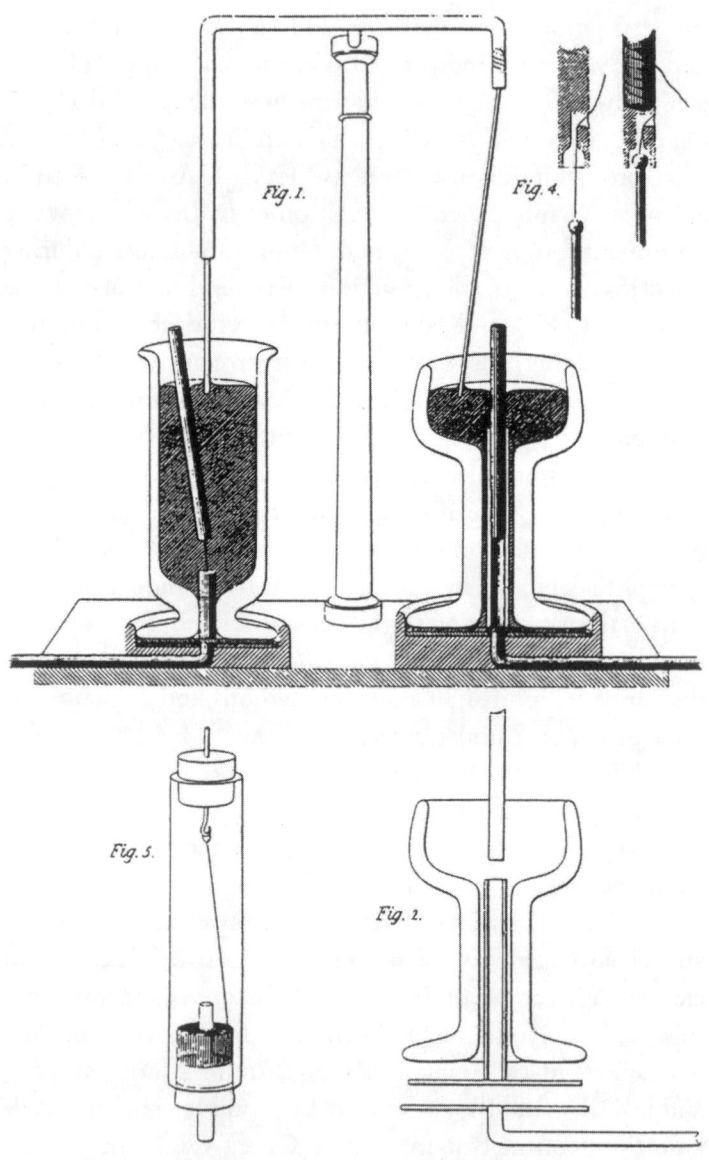

Figure 14. Faraday's apparatus for the measurement of
electromagnetic rotation (1844).

cesco Zantedeschi in Venice had preceded both of them in that discovery. What distinguished Faraday from the others was his continuous productivity in this field, his formulation of general principles of electrochemistry, and his advocacy of the idea of lines of electrical and magnetic force. After his ingenious demonstrations of electromagnetic induction in 1831,[9] Faraday turned, during 1832-1834, to the quantitative comparison of the amount of electricity consumed and the amount of chemical decomposed in electrolysis, and established the validity of two principles: (1) "The chemical power of a current of electricity is in direct proportion to the absolute quantity of electricity which passes"; and (2) "Electrochemical equivalents coincide and are the same with ordinary chemical equivalents."[10] These two principles, which have come to be known as "Faraday's laws," were arrived at experimentally, without the use of mathematics, but born out of the belief in the unity of nature and the interconvertibility of its forces. Moreover, as a chemist, Faraday astutely demonstrated that the chemical action occurs in the solution, not at a distance but at the poles (which he renamed "electrodes'), by passing the electric current though absorbent paper soaked with potassium iodide and starch, and observing the release of iodine, a substance he had come to know when he assisted Davy in experiments during their stay in Paris. This method was later used to good advantage by Joule. The new scientific terms Faraday used in his electrochemical writings—ion, anion, cation, anode, cathode, electrode, electrolyte—were chosen on the advice of the noted polymath William Whewell, at Trinity College Cambridge, and Whitlock Nicholl, a London friend and physician.[11]

During 1833-1837, in addition to his experiments on electrolysis, Faraday examined the phenomenon discovered by Davy and studied by Döbereiner, namely the action of platinum in promoting the combustion of gases, and which Berzelius had included among the examples of catalysis. Faraday questioned Döbereiner's proposed mechanism of the process, but could not offer a satisfactory theory of his own.[12] Of greater consequence were Faraday's studies on the effect of insulators on the distribution of an electric field. He concluded that they store electric charge, and he introduced the term "specific inductive capacity" of such "dielectrics" to denote what was later named "dielectric constant."[13] A few years later he offered a "speculation" about intimate nature of electrical conduction and nonconduction, and expressed his preference for the "point-atoms" of Bošković.[14]

Faraday's strenuous effort in the laboratory (he often worked there from morning to night), together with his self-imposed lectures at the Royal Institution and such service as that to Trinity House on the illumination of lighthouses, led to a nervous breakdown in 1839. Except for more work on the liquefaction of gases,[15] he spent less time in the laboratory. In 1845, however, there was a renewed burst of experimental activity after William Thomson (who had given Faraday's lines of force a mathematical treatment) suggested that Faraday test the effect of electricity on plane-polarized light. In earlier work, Faraday had not observed a detectable effect, but having shown the convertibility of electric and magnetic force, he now tested the action of a strong magnetic field. This produced a measurable rotation of the plane of polarized light, and he found that the angle of rotation was proportional to the strength of the magnetic force.[16] This achievement was immediately followed by a study of the magnetic properties of various materials, and the establishment of a distinction between the "paramagnetic" ones, which were good conductors of magnetism (and aligned themselves along the lines of magnetic force) and poor ("diamagnetic") conductors, which moved away from the lines of magnetic force.[17]

The series of papers on Faraday's experimental studies in electricity and magnetism ended in 1856, when he was in poor health.[18] He continued to give his Christmas lectures for young people; the most famous of these lectures were those in 1860-1861 on the Chemical History of the Candle.[19] Before he resigned from the Royal Institution in 1862, Faraday published articles on various subjects, including the relation of gravity to electricity, lines of force, and field theory. He lived out his final five years in a house in Hampton Court provided by Queen Victoria.

For many of his biographers, including his scientific friends, Faraday was primarily a brilliant experimenter who speculated naively. Pearce Williams has argued convincingly, however, that his early biographers, including his friend John Tyndall (Faraday's successor at the Royal Institution) did not understand Faraday's theories, and failed to appreciate his disciplined thinking.[20] In his published experimental papers, Faraday stressed the methods and results, and his theoretical discussion was brief and occasionally unclear, because he knew that his non-mathematical presentation of novel ideas was likely to elicit the disapproval of his scientific contemporaries. It was the 24-year-old James Clerk Maxwell, while a Fellow of Trinity College Cambridge, who found inspiration in Faraday's concept of lines of force, and then developed models that led to the mathematical formulation of his famous theory of electromagnetism.[21] A brilliant mathematical physicist and skilled experimenter, Maxwell rose to be the first Cavendish professor at Cambridge, and before his untimely death in 1879 he also made important contributions in thermodynamics, the kinetic theory of gases, statistical mechanics, and much more.[22] Maxwell sent Faraday a copy of his paper, *On Faraday's Lines of Force*, read before the Cambridge Philosophical Society during 1855-1856, in which he wrote:

The methods are generally those suggested by the processes of reasoning which are found in the researches of Faraday, and which, though they have been interpreted mathematically by Prof. Thomson and others, are very generally supposed to be of an indefinite and unmathematical character, when compared with those employed by professed mathematicians. By the method which I adopt, I hope to render it evident that I am not attempting to establish any physical theory of a science in which I have hardly made a single experiment, and that the limit of my design is to shew how, by a strict application of the ideas and methods of Faraday, the connexion of the very different orders of phenomena which he has discovered may be clearly placed before the mathematical mind.[23]

Maxwell received an appreciative reply, but it may be questioned whether Faraday would have welcomed Maxwell's later invocation of an imponderable "luminiferous ether" as the carrier of electric, magnetic, and light waves. In his famous treatise, Maxwell also wrote that Faraday's style "was also a mathematical one, although not exhibited in the conventional form of mathematical symbols."[24]

It should be noted that Faraday's conception of electrolysis as a decomposition effected by an electric current was revised after Clausius suggested in 1857 that solutions of electrolytes contain some free ions. Hittorf's studies on ionic mobility, and Arrhenius's theory of ionization, strengthened the view that the function of the electric current is only to carry the released ions to the electrodes. Another feature of Faraday's thought is his use of "equivalents" as defined by Wollaston in 1813. Faraday does not appear to have been influenced by the developments in organic chemistry around 1850 associated with the names of Frankland and Williamson.

In 1881, Helmholtz delivered a Faraday memorial lecture in which he stated that "now the most startling result of Faraday's law is perhaps this. If we accept the hypothesis that the elementary substances are composed of atoms, we cannot avoid concluding that electricity also, positive as well as negative, is divided into definite elementary portions, which behave like atoms of electricity."[25] At the end of his lecture, Helmholtz remarked:

> I am not sufficiently acquainted with chemistry to be confident that I have given the right interpretation, which Faraday himself would have given, if he had been acquainted with the law of quantivalence. Without the knowledge of this law I do not see how a consistent and comprehensive theory could be established. Faraday did not try to develop a theory of this kind. It is as characteristic of a man of high intellect to see where to avoid going further in his speculations for want of facts, as to see how to proceed when he finds the way open.[26]

Faraday did consider the possible atomistic nature of electricity in a section of his seventh series (1834) entitled, "On the absolute quantity of electricity associated with particles or atoms of matter" (§852), but expressed skepticism about Dalton's atomic theory. Others, notably Wilhelm Weber and Richard Laming, had speculated before 1850 about electrical atoms. After the formulation of Maxwell's electromagnetic theory, opinion swung to the view that electricity is a continuous form of matter, and the later revival of the particulate theory was often attributed to Helmholtz's Faraday lecture. This is only true to the extent that the lecture focused attention on the previously neglected writings of George Stoney, who was influenced by Faraday's studies on electrolysis, and had suggested the idea of electrical atomicity in 1874 (at a meeting of the British Association for the Advancement of Science); in a paper published in 1881, he wrote:

> Nature presents us, in the phenomenon of electrolysis, with a single definite quantity of electricity which is independent of the particular bodies acted on. To make this clear I shall

express "Faraday's law" in the following terms, which, as I shall show, will give it precision, viz.: *For each chemical bond which is ruptured within an electrolyte a certain quantity of electricity traverses the electrolyte, which is the same in all cases.* This definite quantity I shall call E_1. If we make this our unit quantity of electricity, we shall probably have made a very important step in our study of molecular phenomena.[27]

Stoney calculated the charge on E_1 (in electromagnetic units) to be about 10^{-21} emu; later estimates were about 4.5×10^{-20} emu. In 1891, he proposed the term "electron" for this unit. That term was adopted for the sub-atomic "corpuscle" discovered in 1897 by Joseph John Thomson (to distinguish him from William Thomson, Lord Kelvin, I follow the practice of referring to him as J.J.).

After receiving a physics degree in 1876 at Owens College Manchester, where he also learned chemistry from Henry Roscoe and Carl Schorlemmer, J.J. went to study mathematics at Cambridge. He remained there the rest of his life, first as Fellow (1881) and Assistant Lecturer at Trinity College, then as Cavendish Professor of Experimental Physics (1884-1919).[28] During the last decades of the nineteenth century, Maxwell's approach to the explanation of natural phenomena—the use of mechanical models and the formulation of equations—marked the style of research at the Cavendish Laboratory and, like many British physicists of his time, J.J. believed in lines, tubes, and vortices in an ether as suitable models. He differed from other mathematical physicists in his abiding interest in chemical problems. Thus, in a treatise published in 1883, he developed William Thomson's theory of vortex rings and applied that theory to an explanation of valency.[29] Although J.J. later gained a reputation as an experimenter, it appears that he was a markedly deficient laboratory worker, and that the work for which he became famous was done by his assistants (notably Ernest Rutherford, John

Sealy Townsend, Charles Thomas Rees Wilson, and John Zeleny), who "did their best to keep him from touching the apparatus."[30]

After his appointment as Cavendish professor, J.J. embarked on an experimental study of electrical discharges in gases. This line of work began in 1858 with Julius Plücker's discovery of rays emitted at the cathode of an electrical discharge tube at very low pressure. He found that these rays traveled in straight lines, were deflected by a magnetic field, and gave rise to fluorescence upon hitting the wall of the tube. These "cathode rays" did not attract much attention until about 1880, when several investigators, notably William Crookes (who modified Plücker's tube) and Arthur Schuster in England, as well as several physicists in Germany (Eugen Goldstein, Heinrich Hertz, Philipp Lenard, and Eilhard Wiedemann) and Jean Perrin in France, were stimulated by the electrodynamic theories of Maxwell and Helmholtz to examine such electrical discharges more closely.[31] The participants in the study of cathode rays differed in their view about their nature, some holding to Helmholtz's particulate theory of electricity, others to Maxwell's theory of electrical disturbance in the ether. In 1896, however, the theoretical and experimental approach was transformed by Wilhelm Röntgen's discovery of X-rays. Of particular importance was the finding in J.J.'s laboratory that a gas exposed to X-rays becomes ionized and a good conductor of electricity, that cathode rays are deflected by an electric field as well as by a magnetic field and this effect is independent of the nature of the gas[32] (Fig. 15). In April 1897, J.J. presented evidence for the view that cathode rays are small, negatively charged corpuscles with mass about 1/1000 of that of a hydrogen atom, and which are constituents of chemical atoms.[33] Similar claims had been made in 1897 by Walter Kaufmann and Emil Wiechert; the latter claimed that his work "showed

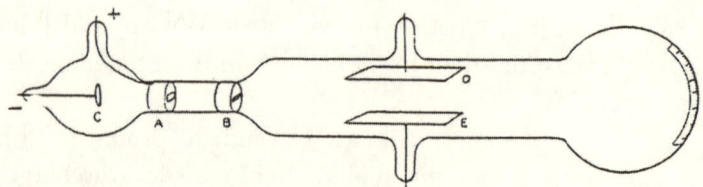

Figure 15. J. J. Thomson's tube for the study of electrial and magnetic deflection of cathode rays (1897).

that we are not dealing with the atoms known from chemistry, because the mass of the moving particles turned out to be 2000-4000 times smaller than the one of the hydrogen atom.[34] Also, early in 1897 a paper by Pieter Zeeman reported that a magnetic field elicited a shift in the wavelength of the spectral lines emitted by incandescent sodium vapor; he estimated that the charge-mass ratio of the "ions" involved is about 1/1000 that of electrolytic hydrogen. This "Zeeman effect" provided experimental evidence for the theory of Hendrik Lorentz that these subatomic ions are in a state of motion around a central nucleus.[35] Lorentz, as well as George FitzGerald and Joseph Larmor, approached the problem from the standpoint of Maxwell's electromagnetic theory of light, and adopted the term "electron" as a substitute for J.J.'s "corpuscle."[36] J.J. did not accept that change until the 1920s. By 1899, J.J. (with Townsend) had estimated the charge and mass separately, but it was not until 1917 that Robert Millikan's famous oil-drop experiments gave values close to those accepted today.[37]

In 1899, J.J. stated that,

I regard the atom as containing a large number of corpuscles. . . . In the normal atom, this assemblage of corpuscles form a system which is chemically neutral. Though the individual corpuscles behave like negative ions, yet when they are assembled in a neutral atom the negative effect is balanced by something which causes the space through which the corpuscles are spread to act as if they had a charge of positive electricity equal in amount to the sum of the negative charges of the corpuscles.[38]

Earlier, Eugen Goldstein and Wilhelm Wien had found that luminous positive rays are generated in gas discharge tubes, and in 1907, the mass-energy ratio for several gases was determined in J.J.'s laboratory.[39]

The elucidation of the nature of the positive rays came largely from the work of Rutherford, who had shown that the rays discovered by Henri Becquerel (Marie and Pierre Curie named the phenomenon "radioactivity") were composed of at least two kinds of radiation, one which was readily absorbed (named α) and a more penetrating one (named β). The β-rays turned out to be the same as cathode rays, while the α-rays were found to be ionized helium atoms. These findings led to the formulation of Rutherford's famous "planetary" model of the atom, with a positively charged nucleus surrounded by negatively charged electrons.[40] During the course of this work, it was found that with neon, two kinds of positively charged particles were found, with masses of 20 and 22. In 1913, Frederick Soddy named elements having the same chemical properties but of different atomic weight "isotopes," and in 1919, Francis William Aston (J.J.'s assistant during 1910-1919), developed the mass spectograph for the identification of isotopes.[41]

After 1900, J.J. turned once again to chemical problems, and in his Silliman lectures at Yale (1903), he presented a model of the atom in which concentric rings of negatively charged corpuscles are imbedded in a positively charged spherical unit. According to this "plum pudding" model, chemical bonds are formed by the transfer of corpuscles from the outer rings of adjacent atoms.[42] He also assumed that the atoms in a molecule are linked by Faraday tubes of force, leading to polarities within the molecule. In subsequent writings, J.J. developed this theme in relation to chemical affinity and a variety of specific topics, such as the Periodic Table, Werner complexes, double bonds, and keto-enol tautomerism.[43]

These writings, which represented a modernized revival of Berzelian dualism, had a considerable impact on chemical thought in the United States and England. In Germany, however, the organic chemists were largely indifferent, and followed Adolf Baeyer and Emil Fischer in the business of the determination of the structure and synthesis of complex compounds, and many physical chemists adopted Wilhelm Ostwald's anti-atomistic energetics.[44] One of the first chemists to apply the electronic theory of matter to the problem of valency was Walther Nernst, who assumed the existence of positive, as well as negative, subatomic particles.[45] He was followed soon afterward by his former assistant, Richard Abegg, who had received his Ph.D. in organic chemistry with August Wilhelm Hofmann. In 1899, Abegg published a paper (jointly with Guido Bodländer) in which he called attention to the correlation of electromotive potentials (as measures of electroaffiinity) of an atom with its position in the periodic table. Abegg then developed this theory further in a subsequent paper: "Every element possesses a positive as well as a negative maximum valence whose sum always is eight. The maximum positive valence is the same as the number of periodic group to which the element belongs."[46] This "rule of eight," which had its origin in earlier formulations of the periodic table, and J.J.'s concept of the alternation of positive and negative charges, provided the basis for further speculations about atomic structure, valency, and the chemical bond. In the United States these ideas were taken up and developed by Kaufman George Falk and John Maurice Nelson, Harry Shipley Fry, William Albert Noyes, and

Julius Stieglitz.[47] In 1916, however, Gilbert Newton Lewis dissented from the prevalent concept of polarized bonds, and proposed the theory that the chemical bond between two atoms consists of a pair of shared electrons that no longer belong to either atom exclusively. Because of the entry of the United States into the war, he was unable to pursue this line for several years, but his full exposition in 1923 rendered the polarity theory superfluous.[48]

A pupil of Theodore William Richards at Harvard (Ph.D. 1899), Lewis bridled at his mentor's emphasis on meticulous experimentation, and discouragement of imaginative thought about atomic and molecular structure. After a year in Germany with Nernst and Ostwald, Lewis returned to Harvard, but left soon afterward, and in 1905 he joined the physical group associated with Arthur Amos Noyes at the Massachusetts Institute of Technology. Seven years later, Lewis became the head of the College of Chemistry at the University of California, Berkeley, where he established one of the leading centers of chemical research.[49]

It is perhaps fair to say that Lewis influenced the development of chemical thought to a greater extent than any other American scientist of his time. Apart from his contributions to the theory of the chemical bond, Lewis was chiefly responsible for the transformation of the chemical thermodynamics of Clausius, Gibbs, and Helmholtz from a doctrine limited to ideal systems to one applicable to real substances and solutions of strong electrolytes. He began this effort as a student, in suggesting the concept he termed "fugacity" (later changed to "activity"), and completed it in his influential book (with Merle Randall).[50] After 1925, he dabbled with relativity theory, but then found a challenge in photochemistry (he introduced the term "photon"), which occupied his attention to his last days.[51]

I return to Lewis's theory of the chemical bond, and its general acceptance by American and British chemists, some of whom began to call themselves "physical-organic" chemists. Of special interest is the advocacy of the Lewis theory by Irving Langmuir. A chemist at the General Electric Company, he worked on many problems of industrial interest, but in 1919-1921, he chose to popularize the theory, and lectured so widely about it that it came to be known as the Lewis-Langmuir theory, or even the Langmuir theory. In 1919, Lewis wrote to Langmuir:

> . . . to be perfectly candid I think there is a chance that the casual reader may make a mistake which I am sure you would be the last to encourage. He might think that you were proposing a theory which in some essential respects differed from my own, or one which was based upon some vague suggestions of mine which had been carefully thought out. . . . It seems to me that the views which I presented were about as definite and concrete as was possible considering the condensed form of publication. I think if any confusion should arise it would be due to points of nomenclature. For example, while I speak of a group of eight you speak of an octet.[52]

Langmuir received the 1932 Nobel Prize in Chemistry for his work (with Katherine Blodgett) on surface films. The prestige of this honor was applied to a dismissal of the peptide theory of protein structure in favor of the so-called "cyclol" theory proposed by the topologist Dorothy Wrinch.[53] It must be added that Lewis was passed over by the Nobel Prize committee, although he was repeatedly nominated between 1922 and 1935 (the last year for which data are available to me), and received eight nominations in 1929.[54]

In 1916, Walther Kossel presented an electronic theory of valence based on the Rutherford-Bohr dynamic model of the atom, with electrons moving in concentric circles about

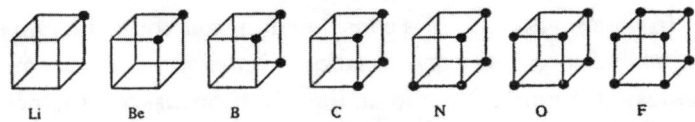

Valence shells of atoms

Iodine molecule   Oxygen molecule

Figure 16. Lewis's cubical models of atoms (1917).

a central nucleus.[55] On the other hand, Lewis used a static cubical model, whose eight corners were sites for occupancy by valence electrons (Fig. 16). In his opinion, this model "presented a picture of a system which, consistently with recognized principles of mechanics and electromagnetics, would give a series of spectral lines analogous to the series which are known for various elements."[56] Kossel's treatment dealt largely with polar (electrovalent) bonds, and he did not propose the shared electron non-polar (covalent) bond, but he also recognized, as had Lewis, that there is a continuum between the two kinds of bonds.

Between 1916 and 1923, when Lewis published his famous book on valence, he had adopted Bohr's model. In that book, he stated that the dualistic theory advanced by Falk and Nelson provided an adequate explanation of the behavior of extremely polar substances, but not of the non-polar compounds of organic chemistry. Lewis also noted that Wendell Latimer and Worth Rodebush added to his theory of valence Maurice Huggins' concept of the "hydrogen bond."[57] Many years later, the hydrogen bond became an important feature of the structure of proteins and nucleic acids.[58]

In the 1923 book, Lewis considered the relation of his concept of valence to the coordination number in Werner complexes, and pointed out that the Co^{3+} ion has no valence electrons, and readily accepts six molecules of substances such as NH_3, which have a lone pair of valence electrons. For the interpretation of the behavior of organic compounds, Lewis provided the important concept of acids and bases as electron-pair donors and electron-pair acceptors, the idea of the displacement (shift) of an electron pair as a determinant in the outcome of chemical reactions, the nature of double bonds, tautomerism, and the structure of benzene. These ideas provided a stimulus for the emergence of physical-organic chemistry during the 1920s and 1930s, and I will return to that development shortly.

It should be noted that Lewis's theory of the shared electron bond was not regarded highly by the mathematical physicists who were developing the new quantum mechanics. One exception was Walter Heitler who, with Fritz London, in 1927 used Erwin Schrödinger's wave equation to develop a theory of the covalent chemical bond.[59] Some years later, after his departure from Nazi Germany, Heitler published a book entitled *Elementary Wave Mechanics* (1945), in which he wrote:

> Long before wave mechanics was known Lewis put forward a semi-empirical theory according to which the covalent bond between atoms was effected by the formation of pairs of electrons shared by each pair of atoms. We see now that wave mechanics affords a full justification of this picture, and, moreover, gives a precise meaning to these electron pairs: they are pairs of electrons with antiparallel spin.[60]

The Heitler-London treatment of the H_2 molecule was followed by the formulation of the so-called valence-bond theory, based on the assumption that the electron orbitals of the individual atoms are unchanged in the molecule. As applied to organic compounds, this theory was satisfactory with respect to single covalent bonds, but Pauling's introduction of the idea of "resonance" to explain the properties of unsaturated organic compounds or the structure of benzene became a subject of lively discussion.[61] An alternative mathematical procedure, the "molecular orbital" method, was introduced by Friedrich Hund and Robert Mulliken, who considered the molecule as a whole rather than as a composite of individual atoms.[62] In the theory underlying this method, it is assumed that in diatomic molecules there are not only bonding electrons but also anti-bonding electrons that oppose a union of the two atoms, and that there are two kinds of orbitals (the wave functions of single electrons), σ-orbitals (localized molecular orbitals spread over two atoms), and π-orbitals (delocalized molecular orbitals spread over an entire molecule). The first kind corresponds to a chemical bond that can rotate about its axis, the second to a bond whose rotation is hindered. In the molecular orbital approach one is therefore dealing with combinations of different kinds of orbitals that make up the electronic configuration of a molecule (Fig. 17).

The initial reception of the two approaches clearly favored the valence-bond theory, in part because Pauling was a much more effective expositor than Mulliken. Subsequently, chemists began to appreciate the greater merits of the Hund-Mulliken theory, largely through the contributions of Erich Hückel to the elucidation of important organic-chemical problems. A former associate of Peter Debye, with whom he formulated the famous theory (1923) of strong electrolytes, and a brother of the organic chemist Walter Hückel, Erich Hückel adopted the Hund-Mulliken approach, and developed a theory of the structure of unsaturated and aromatic compounds in which the unpaired electron orbitals were spread across the entire molecule. This theory indi-

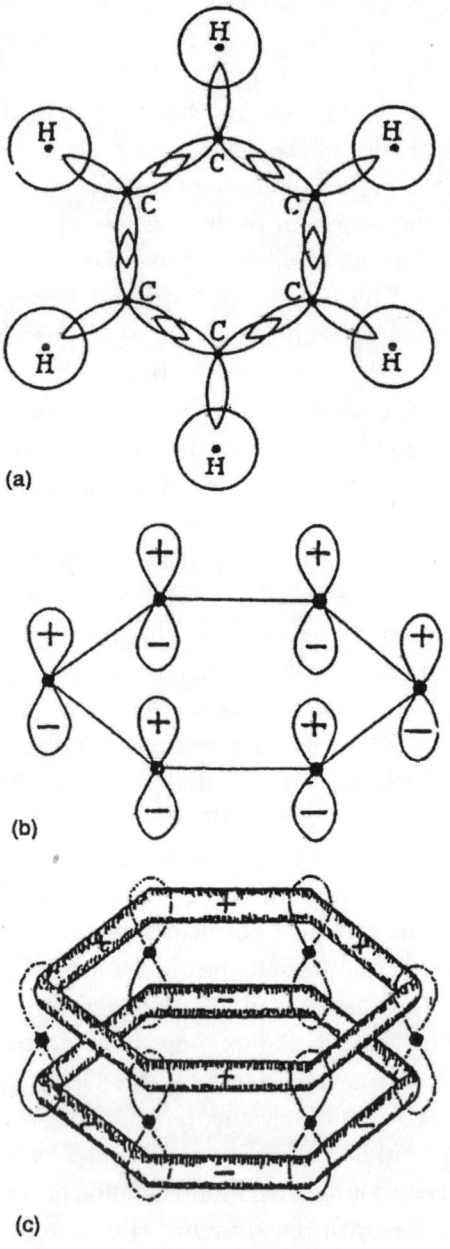

Figure 17. Molecular orbitals of benzene: a. σ orbitals; b. π atomic orbitals; c. π molecular orbitals (Coulson, 1947).

cated that Kekulé had not been far off in suggesting that the benzene molecule is not simply a system of alternating single and double bonds (as he wrote its formula), but one in which there was oscillation of the two kinds of bonds. Moreover, Hückel's treatment gave substance to the theory of Johannes Thiele (1899) of "partial valences" in unsaturated aliphatic compounds. It should be recalled that Thiele explained the finding of Rudolf Fittig and Adolf Baeyer that compounds of the type C=C-C=C added hydrogen at the ends of the chain to form CH-C=C-CH by assuming that such "conjugated" structures contain partial valences.[63] The later development of the molecular orbital method, notably by Charles Coulson, Hugh Christopher Longuet-Higgins, and Michael Dewar made its mathematics somewhat more accessible to chemists.[64] However, the full recognition by synthetic organic chemists of the utility of the molecular orbital method did not come until the 1960s, with the formulation by Roald Hoffmann and Robert Woodward of their "principle of orbital symmetry," which allows one to predict whether a postulated reaction will proceed smoothly.[65] I will return to this important contribution later.

Except for Erich Hückel, most of the prominent theoretical physicists appear to have lost interest in chemical problems after about 1930. The experimental physicists, however, made decisive contributions to chemical research. Among them was Gerhard Herzberg, who provided precise data on the molecular spectroscopy of diatomic molecules.[66] During the 1930s, Isidor Isaac Rabi laid the groundwork for the later invention, by Felix Bloch and Edward Purcell, of nuclear magnetic resonance (NMR) spectroscopy.[67] Together with mass spectrometry, this method became the most valuable analytical technique in organic chemistry; before then, infrared spectroscopy was widely used in organic chemical laboratories.[68] It should be noted that a commercial ultraviolet spectrophotometer only became available during World War II.

Also, the development of X-ray crystallography, introduced by William Henry Bragg and William Lawrence Bragg, had a great impact on organic chemistry during the 1920s, in the demonstration by Kathleen Lonsdale of the planarity of the benzene ring. A few years later, John Desmond Bernal and Dorothy Crowfoot applied the method to the structure of ergosterol and of insulin, and John Monteath Robertson established the three-dimensional structure of metallo-phthalocyanine [69] (Fig. 18).

To these important contributions to chemical methodology must be added the development of new and improved chromatographic methods for the separation of closely related compounds, and the availability of stable and radioactive isotopes (^{2}H, ^{3}H, ^{13}C, ^{15}N, ^{35}S, etc.) for the study of the course and mechanism of chemical reactions.[70] Moreover, after 1920 the concepts of chemical kinetics were enriched by transition state theory and the use of flash photolysis in the study of fast reactions.[71] Some aspects of these developments will be considered later in this section.

During the first decades of the twentieth century, before the quantum theory began to influence chemical thought, several organic chemists approached problems inherited from their predecessors with the thermodynamic and kinetic concepts of the new physical chemistry. This approach was not welcomed by some of the prominent leaders in synthetic organic chemistry, including Adolf Baeyer, who were accustomed to deriving their theories by inspiration or even, as claimed by Kekulé, in dreams. One of these problems was presented by a dispute around 1890 between Johannes Wislicenus and Arthur Michael regarding the stereochemistry of addition and elimination reactions involving derivatives of the isomeric maleic and fumaric acids. The son of affluent

American parents, Michael studied chemistry in Berlin (A. W. Hofmann), Heidelberg (Bunsen), and Paris (Wurtz), and held professorships at Tufts (1881-1889, 1894-1907) and at Harvard (1912-1936). He conducted most of his chemical research at a private laboratory in Newton Center, Massa-

Figure 18. X-ray electron density projection for nickel phthalocyanine (Robertson, 1937).

chusetts. In 1887, Michael described the addition of acetoacetate to the double bond in unsaturated compounds of the
type $R-CH=CHC_6H_5$ (the so-called Michael reaction). In
the same year, Wislicenus published his famous paper on the
determination of the stereochemistry of unsaturated compounds.[72] Michael proceeded to show the limitations of Wislicenus's dependence on van't Hoff's model of unsaturated
compounds and, in an exchange of papers, it became evident
that Wislicenus could not rebut Michael's experimental evidence.[73] Michael did not offer an alternative theory, and his
thermodynamic argument shed little light on the matter,
although in retrospect he has been credited with an anticipa-
tion of the idea of unstable intermediates in chemical reactions.[74] Also, he proposed a "positive-negative" rule according
to which in propene ($CH_3-CH=CH_2$), the methyl group
makes the central carbon atom more positive than the end
carbon atom. The American organic chemist Treat Johnson,
wrote that Michael's

> . . . interpretation of the mechanism of an organic reaction
> as proceeding according to the principle of addition, with
> the formation of an intermediate "addition product," [was
> a] conception which [has] exercised a tremendous influence
> on the development of chemical thought. . . . He has formu-
> lated principles, made and originated syntheses and discov-
> eries leading to applications which have rendered the world
> a service greater than is generally known. . . . [H]e has also
> made most important contributions of theoretical bearing
> dealing directly with such subjects as valency, the relation
> of thermochemistry to organic structure, desmotropism, ste-
> reoisomerism, the theory of partition principle, and the scale
> of combined influence. In the field of organic synthesis may
> be recorded as outstanding accomplishments his original
> method of building up carbon combinations by the direct
> addition of malonic esters, in the form of their sodium salts,
> to unsaturated carbon compounds.[75]

An even less appreciated American pioneer of physical-organic chemistry was Solomon Farley Acree, who published many papers in the field while a member of the chemistry faculty at Johns Hopkins (1906-1914), and who later worked at the National Bureau of Standards.[76] Two better known American organic chemists who contributed to the development of physical-organic chemistry during the first two decades of the twentieth century were John Ulric Nef and Julius Stieglitz, both at the University of Chicago.[77] Both proposed the formation of reactive intermediates in rearrangement reactions; Nef favored bivalent carbon (methylene) compounds (he was severely criticized by Michael), and Stieglitz suggested univalent nitrogen intermediates in the Beckmann (and other) rearrangements of nitrogen compounds. As will be seen later, after 1920, the United States became a major center in the rise in academic status of physical-organic chemistry. It was otherwise in France where, after the death of Wurtz in 1884, the baneful influence of Marcelin Berthelot interrupted the further development of theoretical organic chemistry. Except for some contributions by men such as Marc Tiffeneau, work in this field was not taken up until after World War I, and was markedly idiosyncratic and parochial in its approach.[78]

In England, just before the impact of the discovery of the electron began to influence organic chemical thought, John Norman Collie, professor of chemistry at University College London (1902-1928), and a celebrated mountaineer, studied the constitution of the γ-pyrones obtained from dehydroacetic acid, and suggested that they are oxonium oxides containing a quadrivalent oxygen. He also proposed that the polymerization of the $-CH_2-CO-$ group might lead to the formation of various natural products.[79] Collie's formula for γ-pyrone was later rewritten by Fritz Arndt as a dipolar ion (*Zwitterion*) in which one oxygen is positively charged and the other is negatively charged.[80] The concept of dipolar ions had been applied earlier by Niels Bjerrum, who showed that amino acids have the structure $^+NH_3\text{-}CH(R)\text{-}COO^-$.[81]

Collie's contributions to theoretical chemistry were soon overshadowed by those of two other British chemists, Arthur Lapworth and Thomas Martin Lowry, both pupils of Henry Armstrong in London. Lapworth showed himself to be an excellent experimenter in his Ph.D. work on the sulfonation of naphthol derivatives (1895), and in his collaboration with Frederick Kipping in studies on camphor (1896-1897) and with Collie (1897) on picolines. In 1898, Lapworth revealed his early theoretical bent by seeking "simple laws" for organic-chemical reactions on the basis of stereochemical considerations. Three years later he changed his mind, and concluded that "It is to electrolytic dissociation, often doubtless in extremely minute amount, that the majority of changes in organic compounds may be most probably be assigned."[82] After a few years as head of the chemistry department at the Goldsmith's Institute in London, Lapworth went to Manchester in 1909; he succeeded W. H. Perkin, Jr. as professor of organic chemistry and he remained until his retirement in 1935.[83]

During the course of his research, in addition to further work on camphor derivatives, Lapworth made important contributions to the understanding of the mechanism of reactions of ketones with cyanide (the cyanhydrin reaction), the reaction of compounds having "activated" methylene groups (as in acetoacetic esters) in the condensation reactions associated with the names of Ludwig Claisen and Walter Dieckmann, the phenomenon of tautomerism, and the concept of acids as hydrogen donors in acid-catalyzed reactions (a theory later further developed by Lowry and by Johannes Nikolaus Brønsted). In approaching these problems, Lapworth was guided by the idea of alternating polarities as in ($\overset{-}{C}H_2\overset{+}{C}H=\overset{+}{C}H-\overset{-}{C}H\overset{-}{O}$), but distinguished this notation from that used previously by Fry:

> . . . in attaching the - and + signs in the oxygen and carbon atoms no hypothesis is invoked, nor is it necessary or even desirable to assume that electrical charges are developed on these two atoms (except perhaps at the actual instant of chemical change). The signs are applied, in the first instance, merely as expressing the relative polar characters which the two atoms seem to display at the instant of the chemical change in question.[84]

He named the positively charged atoms as "cationoid" and the negatively ones as "anionoid." After the adoption of the electronic theory of valence, these were recognized to correspond to electron acceptors and electron donors, respectively, and to the terms "electrophilic" and "nucleophilic" later introduced by Christopher Ingold, of whom more shortly.

After receiving his degree in 1896, Lowry was Armstrong's assistant until 1912, when he became lecturer in chemistry at Guy's Hospital; he was promoted to a professorship in 1914. After the war, Lowry was appointed professor of physical chemistry at Cambridge, where he remained the rest of his life. He was deeply religious, and a frequent preacher in the Methodist Church.[85]

Lowry's first research dealt with camphor derivatives, and led him to his two main interests, the mechanism of chemical change and the origin of optical rotatory power. In 1899, he found that halogenation or nitration of camphor gave mixtures of stereoisomeric products, and that the optical activity of a solution of nitrocamphor in an organic solvent changed with time. The latter phenomenon, which he named "mutarotation," had previously been observed with menthone[86] and sugars. Lowry showed that this process was catalyzed by traces of nitrogenous bases, and this finding led him to propose in 1923 that acids and bases be defined as proton acceptors and proton donors (A-H + B = A⁻ + B⁺-H). In the same year, this concept was developed more fully by Johannes Nikolaus Brønsted.[87]

Lowry was an early adherent of Lewis's electronic theory of valence, and applied it to an interpretation of multiple bonds in organic compounds.[88] In his most important paper on this subject, Lowry wrote: "Crotonaldehyde and butadiene are formulated as CH_3-$\overset{+}{C}H$-$\overset{-}{C}H$-$\overset{+}{C}H$-$\overset{-}{O}$ and $\overset{+}{C}H_2$-$\overset{-}{C}H$-$\overset{+}{C}H_2$-$\overset{-}{C}H_2$. It will be seen that, under this scheme, the distinction between single and double bonds in a conjugated system disappears completely. There is therefore no longer any need to postulate a wandering of the double bond when butadiene is brominated, since the central double bond is already in position."[89] Lowry devoted the remainder of his scientific life to the study of optical rotary dispersion and wrote a definitive book on optical activity.[90]

Other British chemists included among the pioneers of the kinetics and mechanism of organic reactions are Nevil Vincent Sidgwick, John Joseph Sudborough, Kennedy Joseph Previté Orton, and Harry Dawson.[91] Before returning to the succession of British and American chemists who created modern physical-organic chemistry, mention should be made of the work of Nikolai Aleksandrovich Menshutkin who, during the period 1876-1907, studied the kinetics of the reaction between tertiary amines and alkyl halides (for example $(C_2H_5)_3N$ and C_2H_5I), and found that the reaction rate depended greatly on the nature of the solvent.[92]

More needs to be said about the work and style of Hans Meerwein, who was successively professor of chemistry at Bonn (1914-1922), Königsberg (1922-1929), and Marburg (1929-1952), and who continued his research after his retirement.[93] In work interrupted by World War I, Meerwein extended the applicability of the Michael reaction to the synthesis of bi- and tricyclic compounds related to camphor. Contrary to Baeyer's strain theory, according to which such saturated ring compounds should be unstable, they were extremely stable, a behavior explained by Sachse's theory of the conformation of cyclohexane, and can exist in two enantiomeric forms ("boat" and "chair"). Meerwein defined the stereochemistry of such compounds, and explained the mechanism of their intramolecular rearrangement, which had been discovered earlier by Egor Egorovich Vagner (Georg Wagner, in the German and English chemical literature), as a process in which there is formed a reactive ionized intermediate containing a "cryptocation" (later named "carbonium" ion). The existence of such ionized carbon compounds had previously been recognized only in the case of the triphenylmethyl radical discovered by Moses Gomberg, and Meerwein's work opened a large area of investigation. Forty years later, George Olah succeeded in preparing the first stable aliphatic carbonium compound. Meerwein also showed that what came to be called the Wagner-Meerwein rearrangement is catalyzed by metal halides, and he explained the catalysis in terms of the intermediate formation of a Werner complex. To these path-breaking contributions, Meerwein added the intermediate formation of positively charged "oxonium" ions in the action of water on ethylene oxides. During the course of these researches, he also introduced a new method for the reduction of aldehydes and ketones, later known as the Meerwein-Ponndorf-Verley method, and, in his final work, he described the reaction of α,β-unsaturated carbonyl compounds with aryldiazonium salts.[94] The originality of these remarkable theoretical and experimental achievements bespeak a style of chemical research rarely matched during the twentieth century. A modest man, and a member of no school, he endured the negative judgment of German journal editors still wedded to the tradition that theoretical interpretations are suspect, and that preference should be given to papers that report experimental results only. More-over, the promises made to Meerwein upon his move to Mar-burg in 1929 were not kept, in part because of the economic depression, the rise of Hitler, and the onset of World War II.[95]

Apart from the later demonstration that, as in the for-mation of reactive carbonium or oxonium intermediates, the reactions of other organic substances, notably the nitro-gen mustard gases [bis (β-chloroethyl)amines], involve the intermediate formation of imonium ions.[96] Meerwein's work also led to the recognition that the racemization of optically active compounds may not involve symmetrical intermedi-ates. Of more general importance was the link he provided between organic chemical mechanism and the thermody-namic interpretation of reaction kinetics introduced by van't Hoff. This approach was developed just before the outbreak of World War I by the brilliant young physical chemist René Marcelin, who was killed in September 1914 in a battle near Verdun.[97] After the war, what came to be called the "transition-state theory" was developed principally by Henry Eyring and Meredith Gwynne Evans, in association with Michael Polanyi. In the present form of the theory, the energy of activation (represented by a peak in a plot of energy versus the reaction coordinate) is related to the rate constant by means of an equation containing the Boltzmann constant, the Planck constant, as well as the enthalpy and entropy of activation. The theory has proved to be a useful qualitative tool in the study of the influence of factors (such as catalysis) on reaction rates.[98]

I return to the British chemists who embraced Lewis's electronic theory of valence. One of them, Robert Robinson, achieved great fame for his work on the synthesis of natural products, but during the 1920s also made significant contri-butions to physical-chemical theory. After his graduate and postdoctoral association with William Henry Perkin, Jr. in Manchester, Robinson was professor at Sydney (1912-1915), Liverpool (1915-1920), St. Andrews (1921-1922), Manchester (1922-1928), University College London (1928-1930), and Oxford (1930-1955).[99]

While in Manchester with Perkin, Robinson worked on the plant dye brazilin, and developed a lasting commitment to the study of the structure and synthesis of natural products, in particular plant pigments and alkaloids. (In later life, he became a passionate collector of exotic flowering plants.) Also, during 1909-1912, he was stimulated by Arthur Lapworth to enter the field of theoretical organic chemistry, and married Gertrude Maud Walsh (a pupil of Chaim Weizmann), who collaborated in many of his later researches and was an enthusiastic fellow mountaineer. The most striking aspect of Robinson's chemical style was his imaginative capacity, based both on a prodigious knowledge of the chemical literature and on immense experience as a synthetic organic chemist, to formulate hypotheses about structures and reactions. He was not dismayed by disproof of the hypotheses that turned out to be wrong, and took delight in the many shown to be fruitful. It does not seem far fetched to connect this capacity to Robinson's considerable skill as a chess player.

Robinson's contributions to theoretical chemistry were made during a ten-year period (1922-1932), and have tended to be overshadowed by his continued massive effort in synthetic organic chemistry and by the rise to prominence of the highly productive physical-organic chemist Christopher Kelk Ingold, of whom more shortly. In a paper (with William Ogilvy Kermack), and subsequent papers, Robinson developed a modification of Lapworth's theory of alternating polarity in terms of Lewis's electronic theory of valence, and applied it to substitution reactions in aromatic and conjugated unsaturated aliphatic systems. He also recognized that organic chemical reactions are ionic in nature, and involve displacement of electron pairs. To indicate such displacement, Robinson introduced the use of curved arrows, a notation that is still in use.[100] He took a dim view of the quantum mechanical approaches to the problem of valence, both of valence-bond and resonance variety and the molecular-orbital method. It has been reported that Robinson found that one of Erich Hückel's predictions was inconsistent with the experimental data, and that it was due to Hückel's mathematical error.[101]

Robinson's first great success came with his synthesis, in one experimental step, of tropinone from succinaldehyde, acetone dicarboxylic acid, and methylamine in water at neutral pH.[102] The previous synthesis by Richard Willstätter in 1901 had required a 20-step process. Over the years, Robinson's group worked on various other groups of alkaloids (isoquinoline, morphine, physostigmine, strychnine), as well as anthocyanins, flavones, steroids, and penicillin. Initially, he claimed that his chemical synthesis of tropinone mimicked the biosynthesis of this compound in the plant, and this view was espoused by others, notably Clemens Schöpf and Arthur Birch, but after radioisotopes became available for metabolic studies, most of the chemical syntheses "under physiological conditions" did not correspond to the reaction pathways in plants.[103]

Robinson's propensity for biochemical speculation was in the tradition of Baeyer's hypothesis that plants make glucose by the polymerization of formaldehyde, or that of Collie, who suggested that the $-CH_2CO-$ group is polymerized to polyketides that are then converted to various natural products. Such speculations are remembered only when they have been shown by experiment to be correct; this was the case in Robinson's use of carbonium ion cyclization to predict the mode of the conversion of the linear squalene molecule into the tetracyclic steroid nucleus.[104]

Other examples of the fruitfulness of such imaginative theoretical contributions to the study of the structure and biosynthesis of natural products can be cited, but few outshine those provided by the work and thought of Leopold Ruzicka, and his younger colleague Vladimir Prelog at the Eidgenössische Technische Hochschule (ETH) in Zürich. Ruzicka demonstrated the validity of the "isoprene [$CH_2=C(CH_3)-CH=CH_2$] rule" for the biosynthesis of linear polyterpenes, and explained the mechanism of their cyclization.[105] In addition to his notable synthetic achievements, Prelog has added a new dimension to stereochemistry through his studies on the conformation of polycyclic compounds, as has Derek Barton, both influenced by the earlier work of Odd Hassel on the electron diffraction of cyclohexane derivatives in the vapor phase.[106]

With Robinson's withdrawal during the early 1930s from active participation in the discussion of organic chemical theory, the leading British laboratory in the field became that of Christopher Kelk Ingold, whose approach was that of a physical chemist rather than that of a synthetic organic chemist. Ingold emphasized the use of kinetic measurements in the study of reaction mechanisms. This approach had a tradition based on the nineteenth-century work of Augustus George Vernon Harcourt, and initiated at Bangor by Kennedy Joseph Previté Orton, who had been greatly influenced by Arthur Lapworth. Orton's kinetic studies were continued at Cardiff by Herbert Ben Watson.[107] One of Orton's pupils was Edward David Hughes, later a close associate of Ingold.[108]

As an undergraduate at University College Southampton (B.Sc. 1913), Ingold acquired a keen interest in physics and mathematics, but switched to chemistry because he found it to be a more challenging field. He did his graduate work at Imperial College London with Jocelyn Field Thorpe. After a two-year stint in industry, Ingold returned to Thorpe's laboratory, received his D.Sc. in 1921, and his massive production of papers earned him election to the Royal Society in 1924. In that year, he succeeded Julius Berend Cohen as professor of organic chemistry at Leeds, and in 1930 he moved to University College London, where he remained until his retirement in 1961. The list of his scientific publications numbers 447 titles.[109]

Up to the mid-1920s, electrons did not figure in Ingold's numerous papers on the mechanism of organic chemical reactions, and he adhered to the view of Bernard Flürscheim that "atoms are not linked by the interposition of electrons."[110] It was only after Lapworth had criticized him in 1924, and Robinson had presented his theory in 1925, that Ingold adopted the electronic theory of valence, and proceeded to build a new "system" of organic chemistry by introducing a new terminology.[111] In 1934, Ingold presented a full-blown summary of his system, in which he used, with only scant acknowledgment, the contributions of others, for example, the curved arrows of Robinson and the replacement of Lapworth's terms anionoid and cationoid by nucleophilic and electrophilic.[112] He also denoted reactions as S_N1 or S_N2 (substitution, nucleophilic, first- or second-order) or E1 (elimination). Ingold has been reported to have said at about that time that "the new work made it inescapably clear that the old order in organic chemistry was changing, the art of the subject diminishing, its science increasing: no longer could one just mix things: sophistication in physical chemistry was the base from which all chemists, including the organic chemist, must start."[113]

Ingold's 1934 review article attracted wide attention, and his later book strengthened the opinion that he had introduced the electronic theory into organic chemistry.[114] As Robinson's biographers put it:

> The belief that the electronic theory was due to Ingold, rather than Robinson, gained strength, at any rate outside Britain, as a result of their very different attitudes to it. To Robinson, his theory was a practical basis for work in the classical areas of organic chemistry, i.e. synthesis and the elucidation of the structures of natural products. During the development of the theory, he carried out extensive experimental investigations of various reactions to help in establishing its principles. Once the theory was fully developed, he virtually abandoned further work in the area, feeling it would be a waste of time since it could at best confirm what he felt were established conclusions. Ingold, on the other hand, as a result of his association with Hughes, became almost wholly absorbed in the study of organic reaction mechanisms and his publications not only made him the acknowledged leader in this field but also kept his theoretical ideas continuously on display.[115]

Over the years, the Ingold group conducted extensive studies on the mechanism of various kinds of organic rearrangements, tautomerism, and substitution, addition, and elimination reactions. Of special interest was work, largely conducted by Hughes, on the mechanism of the inversion of configuration discovered by Paul Walden in 1896. This so-called Walden inversion was the subject of a lengthy investigation by Emil Fischer, who had encountered it in connection with his preparation of α-amino acids from optically active α-bromo acids. In 1907, he offered the speculation that the asymmetric carbon of an optically active compound is attacked by a reagent to form an addition complex, which may decompose with the departure of one of the groups already bound to the asymmetric carbon atom.[116] In his 1923 book, Gilbert Lewis had suggested a backside attack, with the simultaneous departure of one of the previously bound groups, and in 1930 the work of Joseph Kenyon and Henry Phillips indicated that the optical inversion is associated with a bond-breaking step involving the asymmetric carbon atom.[117] In an elegant experiment, using radioactive iodide (I^*), Hughes compared the rate of iodine exchange in (+) $C_6H_{13}CH(I)CH_3$ and (-)$C_6H_{13}CH(I^*)CH_3$ with the rate of bimolecular racemization, and concluded that every inversion of configuration involves a S_N2 substitution, whereas in a S_N1 process a racemic product is produced by an attack of the nucleophile on either side of the planar carbonium ion intermediate.[118] This generalization was challenged by Louis Hammett and Saul Winstein, of whom more later.

It is noteworthy that Ingold's wide-ranging program of research did not include the study of free radicals. A British chemist who considered the possibility that free radicals are intermediates in organic reactions in solution, and who presented experimental evidence in support of this view, was William Alexander Waters. A Cambridge graduate (Ph.D. 1927), Waters worked and taught at Durham (1927-1940) and Oxford (Balliol College, 1944-1971). In his final years he wrote about the history of physical organic chemistry, and deplored "the acceptance too readily of statements which have passed on from being theories put forward in journals to become dogmatic assertions in textbooks. This I think takes little more than 10 years. Few newcomers to research set out to examine critically the work of leading authorities of the previous generation. . . . I have a fear that many more serious errors in chemistry will persist now that our records are being computerized, with the obliteration of contemporary quibbles and cautionary remarks."[119]

Between the two World Wars, Lewis's electronic theory of valence (as expounded by Langmuir) was received more warmly in Britain than in the United States. This difference may be attributed to the aversion to theoretical speculation instilled into leading American professors of organic chemistry by their training in Germany. In many university chemistry buildings, there was an invisible but effective line separating the organic chemists from the physical chemists. There were some exceptions, notably at Harvard, where Arthur Michael and Elmer Peter Kohler influenced the younger James Bryant Conant, and Conant's pupil, Paul Doughty Bartlett. Other leading American pioneers were Louis Plack Hammett (Columbia), Morris Selig Kharasch (Chicago), Howard Johnson Lucas (CalTech), and James Flack Norris (M.I.T.).[120]

In his autobiography, Conant wrote: "I worked so closely with Kohler as a research student, a teaching assistant, and later as a junior colleague, that I am sure that many of my attitudes and opinions are a consequence of his views."[121] A member of a close-knit Pennsylvania "Dutch" (German-speaking) family, Kohler received his Ph.D. degree in 1892 at Johns Hopkins for organic-chemical work with Ira Remsen, and then taught chemistry at Bryn Mawr until 1912, when he moved to Harvard, where he remained the rest of his life. After initial studies on sulfur compounds, Kohler worked on conjugated unsaturated systems, and showed that Thiele's rule of 1,4-addition applied to the reaction of Grignard's reagent (phenylmagnesium bromide) with benzalacetophenone (one of Kohler's favorite compounds). Then followed work on cyclopropane derivatives (a part of Conant's 1916 Ph.D. thesis), isoxazoline compounds, and the resolution of tetrasubstituted allenes ($\overset{a}{\underset{b}{C}}=C=\overset{d}{\underset{c}{C}}$) into two enantiomers. Although a skilled chemical experimenter, Kohler was not attracted to the fashionable synthesis of natural products, but sought to understand the mechanism of organic reactions. In his famous course in advanced organic chemistry, Kohler devoted little time to the great achievements in the synthesis of terpenes or alkaloids, and spoke at length only about the synthesis of sugars, because of its importance in the development of stereochemistry. A brilliant lecturer to students, he avoided participation in large scientific meetings, and preferred to spend time away from Harvard—either in his home town with his family or in solitary walks along mountain trails.

Conant entered Harvard College in 1910 with the intention of becoming a physical chemist and doing his graduate work with Theodore William Richards, and part of his Ph.D. thesis (1916) dealt with electrochemical experiments. Having fallen under Kohler's spell as an undergraduate, Conant completed his formal chemical education at Harvard in six years, with an excellent preparation in both physical and organic chemistry. He returned to Harvard in 1919 to be an assistant professor of chemistry, married Richards's daughter in 1921, and 12 years later was appointed president of the university. In the intervening years, Conant made numerous important contributions in physical organic chemistry. They included the studies of the oxidation-reduction potentials of quinone-hydroquinone systems, of the influence of acidity on the rate of the coupling of diazonium salts with phenols, and of the formation of semicarbazones. In a lecture summarizing the results of such studies, Conant stated:

> There seems no escape from the conclusion that significant comparisons of quantitative measurements can be made only on the basis of a thorough knowledge of a reaction. This requires first a detailed study of all the products and later a physico-chemical study of the factors which control the equilibrium and rate. The amount of work involved in such studies and the complications already unearthed are welcome guarantees that there will be many problems to solve for a long time to come. We may rest confident, moreover, that the fascinating art of organic chemistry will yield only slowly to the devastating inroads of an exact science.[122]

During the 1920s, Conant made an important biochemical contribution by showing that the iron of methemoglobin is the ferric state, and also began work on the determination of the structure of chlorophyll.

As President of Harvard, Conant initiated changes in the undergraduate course of study and in the procedures for the appointment and promotion of faculty members. Upon the outbreak of World War II, Conant vigorously opposed the isolationist movement in the United States, and together with Vannevar Bush (President, Carnegie Institution of Washington), initiated steps toward the establishment of what came to be called the National Defense Research Committee. In 1941, Conant was appointed chairman of the committee and became responsible for the supervision of the uranium fission program. After the war, Conant was a member of an advisory committee of the Atomic Energy Commission that unanimously opposed the development of the hydrogen bomb, and strongly defended Robert Oppenheimer against charges of disloyalty. In 1952, Conant was named High Commissioner of the U.S. Zone of Occupation in Germany, and later became Ambassador to the Federal Republic of Germany, but resigned from the foreign service in 1957, after receiving little help from the State Department to counter the attacks of Senator Joseph McCarthy. Upon his return to the United States, Conant undertook a study of American high schools, and wrote several books on the subject.[123]

Some insight into Conant's style as a scientific investigator is provided by his interest in the history of science, an aspect of his academic career not mentioned by his chemical biographers. Before leaving Harvard for Germany, he set up a program in which undergraduates majoring in the humanities or social sciences had the opportunity to examine several case studies in experimental science. In the introduction to the first of the booklets in the series, Conant wrote:

The cases presented in this series emphasize the prime importance of a broad working hypothesis that eventually becomes a new conceptual scheme. They illustrate the variety of mental processes by which the pioneers of science developed their new ideas. But it must be emphasized that great working hypotheses have in the past often originated in the minds of these scientific pioneers as a result of mental processes that can best be described by such words as "inspired guess," "intuitive hunch," or "brilliant flash of imagination." The origins of the working hypotheses are to be found almost without exception in previous speculative ideas or in previously known observations or experimental results. Only rarely, however, do these broad working hypotheses seem to have been the product of a careful examination of all the facts and a logical analysis of various ways of formulating a new principle.[124]

It should be noted that one of Conant's assistants in this enterprise was Thomas S. Kuhn, later a famous historian and philosopher of science.

Louis Hammett was an undergraduate at Harvard when Conant was a graduate student and teaching assistant there; he also was greatly stimulated by Kohler's lectures. After his graduation in 1916, Hammett spent a year in Zürich with Hermann Staudinger, performed war service, and worked briefly in industry before becoming an instructor in chemistry at Columbia in 1920. He remained there the rest of his active life. In 1923 he received his Ph.D. for work with Hal Beans on the hydrogen electrode, and rose through the ranks to a full professorship in 1935. In those days, the Columbia department was rather weak in organic-chemical research, although instruction in that subject was well presented by John Maurice Nelson (known as "Pop Nelson"), who worked on enzymes. The dominant figure, however, was Marston Taylor Bogert, known as "Colonel Bogert," his rank during World War I, who was interested in perfumes, and whose main activity appeared to be service as an officer of scientific organizations. The head of the department was Henry Clapp Sherman, a nutritionist. Except for Harold Urey's discovery of deuterium and Hammett's physical-chemical studies, there was little to excite a Columbia undergraduate (such as myself) who hoped to become an organic chemist. It was not until 1951, when Hammett became chairman, that new appointments were made that launched Columbia as an important center of organic-chemical research.[125]

During the 1920s, Hammett was strongly influenced by the work of J. N. Brønsted and of Arthur Hantzsch [126] on the nature of acids. As he later wrote:

> By 1927 I was satisfied in my own mind that (1) water is a base in the same sense that ammonia is a base, (2) the so-called hydrogen ion in water solution is OH_3^+ just as it is NH_4^+ in liquid ammonia; (3) measured by any homogeneous equilibrium or rate phenomena HCl is more acid in benzene than it is in water; (4) since water is a base it masks or levels the differences in strength of strong acids, and prevents any determination in aqueous solution of the relative strengths of weak bases. Consequently the study of strong acids or weak bases requires the use of solvents less basic than water; (5) acid indicators and basic indicators are differently affected by a change in solvent.[127]

In 1932, Hammett and his student, Alden Deyrup, had measured the acidity of systems ranging from water to 100% sulfuric acid, using a series of indicators such as p-nitroaniline, and had set up an acidity scale based on the acidity function H_0.[128] In these experiments they used colored indicators and a colorimeter, but the series was later extended when an ultraviolet spectrophotometer became available. Hammett also demonstrated the relationship between the acidity function and the rates of acid-catalyzed reactions, and expanded the concept of basicity to superbasic solutions.

The importance of these contributions cannot be exaggerated, but they were soon over-shadowed by Hammett's equation for the correlation of the equilibria and the rates of a variety of the reactions of ortho- and para-substituted benzoic acids. It states that the logarithms of the rate constants for such reactions are linearly related to the ionization constants of the acids. It is an enormous expansion of the equation proposed by Brønsted and Pedersen for the linear relationship of the logarithms of the rate constants of general acid- or base-catalyzed reactions acidity constants of the acidic or basic catalyst.[129] Hammett introduced two constants: σ to denote the ratio $\log K_s$ (the ionization constant of the substituted benzoic acid)/$\log K_o$ (the ionization constant of benzoic acid itself), and ρ, which denotes the ratio of the corresponding $\log k$ values of the reaction rates. Although the equation $\log k_s/k_o = \sigma\rho$ is an empirical approximation, it is consistent with van't Hoff's thermodynamic formulation of equilibria and reaction rates, and has also been shown to be consistent with molecular orbital theory.[130]

Hammett's treatise, published in 1940, appears to have been the first book bearing the title, *Physical Organic Chemistry*.[131] It greatly stimulated the development of theoretical chemistry, especially in the United States, and its superiority to other books in this field was widely acknowledged.

Before considering the styles of some of the leading American successors of Conant and Hammett, the distinguished pre-electronic work of Moses Gomberg (Michigan) should be recalled. He discovered the trivalent triphenylmethyl free radical, and opened the field subsequently developed by Waters and others.[132] Gomberg's work was criticized by James Flack Norris (a pupil of Ira Remsen) who later did valuable research on the relative rates of the reaction of alcohols and acyl halides and on the Friedel-Crafts reaction.[133] Among the older organic chemists who readily adopted the electronic theory of valence was Morris Kharasch (Chicago).

He made important contributions to the study of organic reaction mechanisms, notably in the discovery of the "peroxide effect" in the reaction of unsaturated compounds with HBr.[134]

Another early adherent of the electronic theory was Howard Lucas (CalTech), an excellent experimenter and outstanding teacher; one of his students was Saul Winstein, with whom he published a paper correcting the Hughes-Ingold mechanism of the Walden inversion.[135]

At the end of World War II, the graduate students and postdoctoral associates of Kohler, Conant, Hammett, and these other American organic chemists, not only developed the discipline of physical organic chemistry, but also demonstrated how its theoretical principles and those of stereochemistry could be applied fruitfully in the synthesis of complex natural products, and how organic synthesis could provide new compounds whose properties required the formulation of new or revised principles. Among these younger men were Paul Bartlett, Saul Winstein, and Robert Woodward. An examination of their styles of research may provide a glimpse of the conceptual structure and practical methodology of organic chemistry in the recent past.

After completing his undergraduate studies at Amherst (1928), Bartlett received his Harvard Ph.D. degree (1931) for work with Conant on semicarbazones. Although Bartlett's promise was recognized, he was not offered a faculty post at Harvard, it being a departmental policy to deny such appointments to their own graduates. After a three-year "exile" at the Rockefeller Institute (1931-1932) and Minnesota (1932-1934), he was granted the Harvard instructorship, and rose through the ranks to become a full professor in 1946. He retired in 1975, moved to Texas Christian University (Fort Worth) and back to Harvard in 1985.[136] After 1934, Bartlett explored various organic chemical problems, and in 1939 won great acclaim for his paper (with Lawrence Knox) on the synthesis of 1-bromonorbornane and the demonstration that the bridgehead halogen is chemically resistant. Bartlett recognized that this stability is a consequence of the non-planar nature of the bicyclic compound, whereas a compound such as tertiary butyl bromide can readily ionize to form a planar carbonium ion that can be attacked by a nucleophilic reagent.[137] Bartlett continued this line of work after World War II; during the war years, he studied the hydrolysis of nitrogen mustard gases, and demonstrated the intermediate formation of imonium ions.[138] Other main areas of his research dealt with the catalysis by aluminum chloride of the halogen-hydrogen exchange organic halides and isoparaffins, with α- and β-lactones, with the synthetic use of alkyllithiums, with cycloadditions, with the mechanism of the diazo coupling of aromatic amines, and with kinetic studies on the polymerization of free radicals. It may perhaps be too soon to pass judgment, but in the opinion of John Roberts (CalTech; postdoctoral fellow at Harvard 1948-1949):

> I am confident that Paul Bartlett will be recorded by history as the man who was more responsible for the vitality of American physical organic chemistry than any other person. Great credit is also due Louis Hammett of Columbia University, whose early book was tremendously influential. However, Bartlett went much further than Hammett through uniting organic and physical chemistry—synthesis, structure, and mechanisms—whereas Hammett's contribution was more that of a physical chemist interested in organic reactions. How did Bartlett compare with Ingold, who many may think of as the quintessential physical organic chemist? For one thing, Ingold did not leave a great legacy of skilled practitioners in the field. There is clearly a Bartlett school of physical organic chemists, really no longer a viable Ingold school. Perhaps the reason is personality, or perhaps

a more likely reason is that Ingold contributed mostly to the study of a very limited set of important mechanisms and far less to organic chemistry as a whole.[139]

In addition to Roberts, Bartlett's postdoctoral associates included Saul Winstein (1939-1940), who received his undergraduate degree (1934) at the University of California, Los Angeles (UCLA), and his Ph.D. (1938) at CalTech, where he worked with William Young and Howard Lucas. Two years later, Winstein returned to UCLA as an instructor, and became a full professor in 1947. He died of a heart attack in 1969, at the height of his brilliant career.[140]

During the course of a prodigious output of chemical papers, Winstein greatly extended the concepts proposed by Hammett and Ingold regarding the mechanisms of organic reactions. In an extensive series of experimental studies, beginning with his work as a student, Winstein showed that, in addition to the participation of the solvent ("solvolysis") and the electronic displacements in S_N1 or S_N2 reactions, an important effect was exerted by neighboring substituent groups; he named this effect "anchimeric assistance." For example, he showed that the 5,6-double bond in cholesteryl *p*-toluenesulfonate participates in the solvolysis of this compound,[141] with the formation of an intermediate (transition state) cation in which the charge is spread ("delocalized") over several carbon atoms to form a "homoallylic ion" or "nonclassical ion." The product of this reaction was an isomeric cholesteryl derivative, with the substituent at carbon 6 (instead of carbon 3), and the subsequent demonstration of the reversibility of the process indicated the formation of the same intermediate. In later work, Winstein also discovered a salt effect in the ionization process, and examined its relation to the behavior of various kinds of ionic intermediates, from intimate ion pairs and solvent-separated ion pairs to dissociated ion pairs.

The massive experimental evidence that came from Winstein's laboratory, and elsewhere, led to wide acceptance of his theory of neighboring group participation and the intermediate formation of nonclassical ions.[142] Among the skeptics was Herbert Charles Brown, whose objections were rebutted by Winstein.[143] In connection with this work, Winstein displayed great ingenuity, and a commanding knowledge of organic chemical methodology, in the synthesis of many new compounds, some of exceptional beauty.[144] Among these compounds were "homoaromatic" ions of enhanced stability, owing to the replacement of a double bond by a cyclopropane ring. Moreover, he was quick to take advantage of the availability of new or improved physical methods, such as nuclear magnetic resonance spectroscopy or mass spectrometry.

In the absence of a definitive biography, it can only be inferred from the available data that, apart from his family life, Winstein was wholly devoted to the pursuit of his kind of chemistry. Like many of his generation and social class who came of age in the United States during the early 1930s, he probably had to overcome obstacles to the fulfillment of his ambition, and became a tenacious competitor. The recognition he gained was clearly hard-won, through intensive work, deep study, and exceptional inventiveness.

Few chemists have attained the legendary status accorded to Lavoisier, Liebig, Faraday, or Pasteur, but a probable addition to that short list is Robert Burns Woodward. All who knew him were dazzled by his brilliance, and forgave his eccentricities, chief among them an arrogance that masked a generous but lonely spirit. A few of his many graduate students and postdoctoral associates attempted to imitate his style, only to make themselves objectionable.

A wayward child, Woodward's love of chemistry was first

centered on the laboratory he set up at his home in Quincy, Massachusetts. After some argument, he was admitted to the undergraduate program at the Massachusetts Institute of Technology (MIT) in 1933 at the age of 16, and received the B.Sc. degree three years later only after he had reluctantly agreed to meet the physical education requirement in the gymnasium. Woodward's life-long aversion to physical exercise was proverbial. One year after enrolling in the MIT graduate program, he was granted his Ph.D. for work done without the direct supervision of a faculty member. This generous treatment was a consequence of the recognition, by James Flack Norris, of Woodward's exceptional promise.

Later in 1937, after teaching during the summer at Illinois, Woodward returned to Boston, and became an assistant of Kohler, who arranged for him to be elected a member of the Harvard Society of Fellows. Woodward was made an instructor in 1941, and attained a full professorship in 1950. He remained at Harvard until his death in 1979.[145]

Features of Woodward's style of research became evident at the start of his career. While still a member of the Society of Fellows, he published the first of a series of papers on the ultraviolet absorption spectra of unsaturated compounds; at that time the Beckman DU ultraviolet spectrophotomer had just become commercially available. This readiness to use physical instruments for the elucidation of structure and as an aid in synthesis was evident in later years. In 1958, Woodward introduced the "octant rule" for the interpretation of optical rotatory dispersion data. Moreover, infrared spectrometry, mass spectrometry, nuclear magnetic resonance spectrometry, and X-ray crystallography all played a part in Woodward's research, as did the invention of high-performance liquid chromatography for the purification of intermediates and products in multi-step syntheses.

In his first years at Harvard, Woodward approached the problem of steroid synthesis by the use of the Diels-Alder diene reaction, which continued to occupy his attention in later years. He also began to use his encyclopedic knowledge (and phenomenal memory) of chemical reactions to plan possible routes for the synthesis of natural products such as quinine and cholesterol. Years later, Woodward wrote: "Synthesis must always be carried out by plan, and the synthetic frontier can be defined only in terms of the degree to which realistic planning is possible, utilizing all of the intellectual and physical tools available."[146] His appointment in 1941 as a member of the Harvard faculty allowed Woodward to accept graduate and postdoctoral students. With the financial support of Edwin Land (Polaroid), Woodward (with William von Eggers Doering) achieved the total synthesis of quinotoxine in 1944, in a carefully-planned but difficult 20-step procedure; it was later shown to be convertible to the antimalarial quinine. This achievement was hailed in the public press. As the size of his research group grew, and Woodward began to travel abroad, he ceased to participate in the laboratory work, and only planned and supervised the labor of his associates. He maintained close relations with several American pharmaceutical companies, and in 1963 he agreed to the establishment by the Ciba (later Ciba-Geigy) company of the Woodward Research Institute next to its headquarters in Basel; there, he directed work on antibiotics and prostaglandins.

After World War II, when Woodward's fame was assured, he received many lecture invitations, and developed a theatrical style of presentation. He invariably wore a blue suit, white shirt, and light blue tie. He disdained the use of lantern slides; instead, he required a large blackboard, and brought a set of colored chalks and a clean eraser. In describing a multi-step synthesis, Woodward began to write, with meticulous accuracy, at the upper left corner of the black-board, and continued downward across the blackboard until he reached, simultaneously, its lower right corner and the end of his lecture. Such showmanship required, of course, much rehearsal. There were ornate phrases, and occasional military metaphors.[147] Like other kinds of great actors, Woodward basked in the enthusiasm of his audience, and his purpose was clearly to demonstrate his ability to separate himself from the image of an ordinary scientist. His passion-ate devotion to chemistry, evident to all who heard his public lectures, carried a heavy price, for his two marriages both ended in divorce, due to the inevitable conflict between his total commitment to research and his obligations as a hus-band and father of four children.[148]

After the work on quinine, Woodward's achievements included the total synthesis of cholesterol and other steroids (1951-1954), strychnine (1954), lysergic acid (1954), chloro-phyll a (1960), colchicine (1963), cephalosporin C (1966), and vitamin B_{12} (1973); the last (Fig. 19) was accomplished in a collaboration with Albert Eschenmoser (ETH, Zürich). In his public lectures, it was Woodward's practice to use one of these syntheses as the main theme. When he received the 1965 Nobel Prize in Chemistry ("for his meritorious contri-butions to the art of organic synthesis"), he described the synthesis of cephalosporin C, performed at his institute in Basel.[149] Before the Harvey Society in New York, he spoke about the synthesis of colchicine.[150]

In the planning and execution of these syntheses, Wood-ward exhibited a remarkable capacity to visualize the three-dimensional structure of the product, to discern the structural elements which might best serve as starting materials, and to select from the vast armamentarium of organic reactions those likely to permit mechanistic and stereochemical control of each step in the synthesis.[151] In connection with some of these syn-

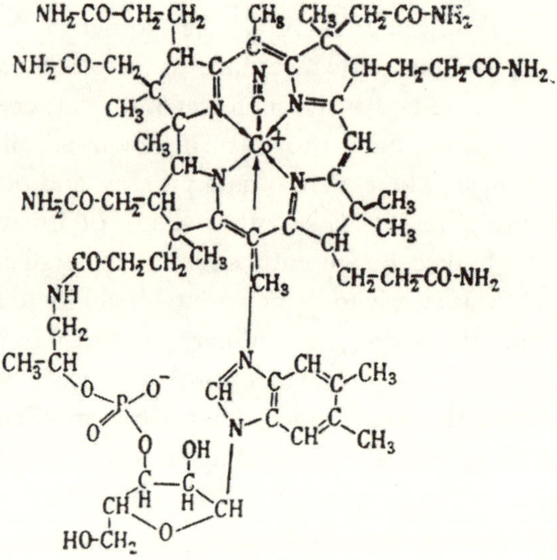

Figure 19. Vitamin B$_{12}$ (Cobalamin).

theses, as in the case of steroids, Woodward was obliged to investigate the validity of proposed mechanisms.[152]

Woodward also made important contributions to the elucidation of the structure of several natural products. In 1944 he offered cogent arguments in favor of the β-lactam structure (as against the thiazolidine-oxazolone structure) of penicillin, which was established by the X-ray crystallographic work of Dorothy Hodgkin. There followed notable contributions to the determination of the structure of strychnine (1947), cevine (1954), and tetrodotoxin (1964). Like Robinson, Woodward speculated about the possible pathways in the biosynthesis of natural products, and also recognized the possible role of squalene as a direct precursor of the steroid nucleus,[153] but other biogenetic predictions were not confirmed in biochemical studies.

In the total synthesis of vitamin B$_{12}$, the unexpected outcome of one of the steps led to an important advance in the understanding of organic-chemical reactions. The empirical discovery was that in the thermal cyclization of a key intermediate, the formation of the expected product was accompanied by the appearance of its diastereoisomer. It was also found that the reverse opening of the newly formed ring by irradiation was stereospecific. To explain these findings, Woodward and Roald Hoffmann (at that time a member of the Harvard Society of Fellows) applied molecular orbital theory to the examination of the symmetry requirements in such reactions, and demonstrated that in order for two atoms to form a bond, the overlapping portions of the two highest molecular orbitals must be of the same sign. In chemical reactions, to achieve this state, molecules undergo rotatory motion around carbon-carbon bonds, thus determining the stereochemical outcome of the reaction. When an electron is promoted by irradiation to an orbital of higher energy (an "excited state"), a previously disrotatory motion may be changed to a conrotatory motion. In their first paper, Woodward and Hoffmann formulated the rules for thermal (ground state) electrocyclic reactions: In systems containing 4q π-electrons, the bonding must involve orbital overlap on opposite faces of the system, and in those containing 4q + 2 π, electrons, the overlap must involve orbital envelopes on the same side of the system. In photochemical processes, the symmetry is different and the stereochemical outcome is reversed.[154] This principle of the control by orbital symmetry of the feasibility and stereochemical consequenes of concerted chemical reactions was further developed by Woodward and Hoffmann,[155] as well as others, notably Luitzen Oosterhoff and Kenichi Fukui, who had independently reached the same conclusions by a different route.[156] The principle quickly became one of the basic tenets of modern organic chemistry, and is a prime example of the emergence of an important scientific theory from the tenacious experimental scrutiny of an unexpected empirical finding.

The art of organic synthesis, and its role in the development of theoretical chemistry, have received continued stimulus from the demands for new medicinal agents. During the final decades of the twentieth century, numerous investigators, most notably Elias James Corey, have followed Woodward in devising elegant procedures for the laboratory synthesis of complex organic compounds.[157]

Chapter Ten

CONCLUSION

⤳

THE RESEARCH STYLES of the chemical investigators considered in this book varied widely and, in some cases, changed markedly during their scientific careers. In their styles of chemical reasoning, they reflected the several categories listed by Crombie, and often combined them with one another and with common-sense thought. There were those who sought to elevate the status of their field to that of the mathematical physics of their time, while others were content to study the individual properties of the growing number of purified substances, and sought schemes for their classification in a manner resembling the taxonomic efforts of contemporary biologists. In their attempts to discern the intimate constitution of chemical substances, some investigators drew inspiration from crystallography, while others relied only on the outcome of systematic studies of chemical reactions and on chemical synthesis. Although many great experimenters proclaimed their distrust of a priori theories, they offered speculations of their own. Some famous leaders of research groups were highly skilled in the conduct of chemical operations, whereas others were markedly deficient in that regard, and depended on the labor of their assistants and research students. There were also differences in the readiness to acknowledge the contributions of such younger people, of predecessors, or of contemporaries.

Broad generalizations drawn from the careers of the chemists considered in this book are therefore likely to be wrong, and I only note the frequency with which experience in early life have formed preconceptions evinced in mature research. One example is the electrochemical work that Berzelius did with Hisinger. Another is the contrast between the approaches of van't Hoff and Nernst to problems in physical chemistry, with van't Hoff entering the field from organic chemistry, and Nernst from experimental physics. For Emil Fischer, it was the discovery of phenylhydrazine, for Woodward, his fascination with the Diels-Alder reaction.

The perceptive reader will have noticed in the sketches of the work and thought of important contributors to the development of chemistry some differences in the extent of my admiration of their methods and styles. In that respect, I often differ from the judgment of professional historians who have specialized in the study of the careers of particular eminent chemists. A notable exception is Faraday, but I am less worshipful of Lavoisier as the revolutionary founder of modern chemistry, of Liebig as the greatest chemist of the nineteenth century, of Pasteur as a chemical crystallographer, or of Kekulé as the formulator of valence and structure theory. In method and style of organic chemical research, I place Wöhler ahead of his renowned friend Liebig, Laurent ahead of his friend Gerhardt, and Wurtz ahead of Dumas. Likewise, in the formulation of the first law of thermodynamics, Joule's style appeals to me more than that of Helmholtz, and I consider the style of Helmholtz's contributions to the development of chemical thermodynamics to be less impressive than those of Clausius, Horstmann, van't Hoff, and Gibbs.

In comparing the methods and styles of pairs of competitors, I have occasionally been influenced by the difference in their social status. For example, despite Priestley's shortcomings as a chemical investigator, and his obstinate refusal to accept Lavoisier's interpretation of the reactions of oxygen and the composition of water, I showed greater sympathy for Priestley than for Lavoisier in their competition during the 1770s. I did so because at that time, the older Priestley was still eking out a living as a Nonconformist minister and then as librarian to Lord Shelburne, while the wealthy Lavoisier was rising rapidly in the Parisian establishment as a member of the Ferme. Priestley's famous meeting in Paris with Lavoisier in 1774, about which there have been different opinions, was made possible by the generosity of Shelburne in taking Priestley as a companion on a trip to the continent. Although I reject the claims for the social construction of scientific knowledge, I believe that during the eighteenth century the social status of scientists played a significant role in their competition for preferment. Some historians may dispute the relevance of such factors and, in their meticulous study of a famous scientist, may prefer to base their interpretation solely on selected items drawn from the archival record. If, in their opinion, my disposition to do otherwise makes for bad history, I regret our disagreement.

During the nineteenth century, with the entry into chemical research of many young people from relatively poor families, social status played a less significant role in professional preferment. Instead, there emerged in France and the German states sizable research groups headed by university professors whose judgment of the merit of a junior associate was frequently based on selfish considerations. Wöhler, Wurtz, and Baeyer were notable exceptions and, as a consequence, they exerted, through their scientific progeny, a greater influence on the development of chemistry. Such large groups did not appear in the United Kingdom until later in the century, but the influx of able young people was spurred by the Industrial Revolution, and numerous chemists of modest social origin were raised to knighthood.

In the twentieth century, with the consequences of its two World Wars and the rise of fascism, vast flows of emigration from continental Europe and later from Asia occurred. Many of the older immigrants had been established scientists in their home countries and enriched the already flourishing chemical research in England and the United States, but many more were people of modest means whose sons and daughters were given the opportunity to become leading chemists. Although some of the unattractive features of German university life, notably the discrimination against women and Jews, was carried over into American universities, such unequal treatment has gradually abated after World War II.

For historians of the twentieth-century sciences, as well as for the educated public, a new factor in the evaluation of the methods and styles of research scientists has been the introduction in 1901 of the Nobel Prizes. That nearly all the awards in chemistry have been hailed within the knowledgeable scientific community attests to the care with which the selections were made, although the omission of some worthy candidates (notably Gilbert N. Lewis) has been widely deplored. The temporary public fame accorded the recipients now represents the principal mark of social status, and reflects the scientific fashions of the time.

In a recent book, some sociologically minded medical historians have introduced new terms to denote the twentieth-century impact of chemistry on the development of biological knowledge and medical practice. They are "molecularizing" and "molecularization." [1] In the introduction, the editors state: "It is our aim in this volume to draw attention to the formation of particular strategic approaches to biology and medicine centered on molecules. These approaches became prominent in the interwar period and gained new momentum with the biomedical mobilization of World War

II. The identification, production, circulation, and uses of molecules in biological research and in the explanation and treatment of diseases created new links between the laboratory, the clinic, and industry. We introduce the term "molecularization" to describe the creation and transformation of these alliances." [2] There follow well-documented chapters, by various authors, on colchicine, on the work of Van Slyke, on vitamins, on Cohn's wartime blood protein program, on Stanley's role in the cancer crusade, on sickle cell hemoglobin, on phenylketonuria, on the oral contraceptive pill, and on the immunotherapy of cancer. A striking feature of these articles, most of which are outstanding in their scientific content and in the appreciation of the social role of chemists such as Van Slyke, Cohn, Stanley, and Pauling, is the seemingly studied avoidance of such terms as "chemistry" or "chemical substance." What pharmaceutical companies produce are stated to be "molecules," and there is also a disregard of the crucial role of synthetic organic chemists in that process. Indeed, it is implied that after 1960, chemistry became another "biomedical science," a term introduced in that year by a representative of the National Institutes of Health. I believe that it is important to trace the social factors in the interactions among laboratory workers, granting agencies, and industrial companies, but it is, in my opinion, a distortion of the historical record to diminish the central role of chemical analysis and synthesis in the development of modern biological knowledge and medical practice. I borrow from sociologists the concept of self-interest in suggesting that the impulse to redefine the role of chemistry as the "molecularization" of the "biomedical sciences" is related to the dependence of the editors and authors of this book on the financial support they receive from medical granting agencies. These historians of "molecular biology" have failed to notice the emergence, during the 1990s, of "chemical biology" as the preferred designation of the research pursued by an increasing number of younger organic chemists interested in biological and medical problems.[3]

NOTES

Foreword

1. R. McKeon (ed.) (1941). *The Basic Works of Aristotle*. p. 474. New York: Basic Books.

2. J. A. Simpson and E.S.C. Weiner (eds.) (1989). Style. *Oxford English Dictionary*. 2nd ed. Vol. 16, pp. 1008-1010. Oxford: Clarendon Press.

3. S. S. Toulmin (1963). Discussion of paper by T. S. Kuhn in *Scientific Change* (A. C. Crombie (ed.), p. 383. London: Heinemann. See T. Nickles (1995). Philosophy of science and history of science. *Osiris* 10:137-163.

4. J.G. McEvoy (1997). Positivism, Whiggism, and the Chemical Revolution: A study in the historiography of chemistry. *Hist. Sci.* 35:1-32. An extreme example of the antipathy of some professional historical specialists is provided by P. Munday (1997). Justus Liebig's research school: Historiographical artifact and anachronism. In *Biology Integrating Scientific Fundamentals* (B. Hoppe, ed.), pp. 398-414. Munich: Institut für Geschichte der Naturwissenschaften.

5. P. Duhem (1947). *The Aim and Structure of Physical Theory* (translated by P. P. Wiener), pp. 69-71. New York: Atheneum; L. Daston and M. Otte (eds.) (1991). Style in science. *Sci. Context* 4:223-447; J. Harwood (1993). *Styles of Scientific Thought: the German Genetics Community, 1900-1993*. University of Chicago Press; M. J. Nye (1993). National styles. French and English chemistry in the nineteenth and early twentieth centuries. *Osiris* [2] 8:30-49.

6. L. Fleck (1979). *Genesis and Development of a Scientific Fact*. University of Chicago Press; K. Mannheim (1952). *Essays on the Sociology of Knowledge*. Oxford University Press.

7. T.S. Kuhn (1962). *The Structure of Scientific Revolutions*. (1977). *The Essential Tension*. University of Chicago Press; (1963). The function of dogma in scientific research. In *Scientific Change* (note 3), pp. 347-364. See J. L. Heilbron (1998). Thomas Samuel Kuhn. *Isis*

89:505-515; P. Horwich (1993). *World Changes: Thomas Kuhn and the Nature of Science*. Cambridge, Mass.: MIT Press; J. Andresen (1999). Crisis and Kuhn. *Isis* 90:543-567.

8. K.L. Caneva (1998). Objectivity, relativism, and the individual: A role for a post-Kuhnian history of science. *Stud. Hist. Phil. Sci.* 29:327-344.

9. P. Geyl (1955). *Use and Abuse of History*. p. 80. New Haven: Yale University Press.

10. A. C. Crombie (1994). *Styles of Scientific Thinking in the European Tradition*. London: Duckworth.

11. A. C. Crombie (1995). Commitments and styles of European scientific thinking. *Hist. Sci.* 33:225-238 (237).

12. M. Bunge (1962). *Intuition and Science*. Englewood Cliffs, N.J.: Prentice-Hall; I. Hacking (1992). 'Style' for historians and philosophers. *Stud. Hist. Phil. Sci.* 23:1-20; M. Vicedo (1995). Scientific styles: Toward some common ground in the history, philosophy, and sociology of science. *Perspect. Sci.* 3:231-254.

13. A. R. Hall (1969). Can the history of science be history? *Brit. J. Hist. Sci.* 4:207-220; H. Kragh (1987). *The Historiography of Science*. Cambridge University Press; T. Nickles (1992). Good science as bad history; from order of knowing to order of being. In *The Social Dimension of Science* (E. McMullin, ed.), pp. 85-129. University of Notre Dame Press; S. Brush (1995). Scientists as historians. *Osiris* 10:215-231.

14. D. L. Hull (1979). In defense of presentism. *Hist. Theory* 18:1-14; A. R. Hall (1983). On Whiggism. *Hist. Sci.* 21:45-59; M. Bunge (1991, 1992). A critical examination of the new sociology of science. *Phil. Soc. Sci.* 21:524-560, 22:46-76.

15. J.A. Fodor (1974). Special sciences (or the disunity of science as a working hypothesis). *Synthese* 28:97-115; J. Dupré (1993). *The Disorder of Things: Metaphysical Foundations of the Disunity of Science*. Cambridge, Mass.: Harvard University Press; P. Galison and D. J. Stump (eds.) *The Disunity of Science*. Stanford University Press.

16. J.J. Berzelius (1813). *A View of the Progress and Present State of Animal Chemistry*. London: Hatchard, Johnson and Bonney.

17. M.J. Nye (1993). *From Chemical Philosophy to Theoretical Chemistry*. Berkeley: University of California Press.

18. J. W. Servos (1990). *Physical Chemistry from Ostwald to Pauling*. Princeton University Press; K. J. Laidler (1993). *The World of Physical Chemistry*. Oxford University Press; E. N. Hiebert

(1996), Discipline identification in chemistry and physics. *Sci. Context* 9:93-119.

19. J. S. Fruton (1999). *Proteins, Enzymes, Genes: The Interplay of Chemistry and Biology.* New Haven: Yale University Press.

20. D. H. Galaty (1974). The philosophical basis of mid-nineteenth century German reductionism. *J. Hist. Med.* 29:295-316; N. L. Maull (1977). Unifying science without reduction. *Stud. Hist. Phil. Sci.* 8:143-162; H. Primas (1983). *Chemistry, Quantum Mechanics and Reductionism.* Berlin: Springer; S. J. Weininger (1984). The molecular structure conundrum: Can classical chemistry be reduced to quantum chemistry? *J. Chem. Educ.* 61:939-944; E. R. Scerri (1993). Correspondence and reduction in chemistry. In *Correspondence, Invariance, and Heuristics* (S. French and H. Kamminga, eds.), pp. 45-64. Dordrecht: Kluwer.

21. M. Friedman (1992). *Kant and the Exact Sciences.* p. 217. Cambridge, Mass.: Harvard University Press.

22. A. Koyré (1958). *From the Closed World to the Infinite Universe.* New York: Harper; H. Metzger (1987). *La Méthode Philosophique en Histoire des Sciences* (G. Freudenthal, ed.). Paris: Fayard. See C. C. Gillispie (1973). Alexandre Koyré. *Dictionary of Scientific Biography* 7:482-490; G. Freudenthal (ed.) (1990). *Études sur Hélène Metzger.* London: Brill.

23. A. Franklin (1986). *The Neglect of Experiment.* Cambridge University Press; A. Pickering (1995). *The Mangle of Practice.* University of Chicago Press; N. Roll-Hansen (1998). Studying natural science without nature? Reflections on the realism of so-called laboratory studies. *Stud. Hist. Phil. Biol.* 29:165-187.

24. R. K. Merton (1959). *Social Theory and Social Structure.* Glencoe: Free Press.

Chapter One

1. E. T. Bell (1933). *Numerology.* Baltimore: Williams & Wilkins; J. Gullberg (1997). *Mathematics.* p. 25. New York: Norton.

2. K. Fritz (1976). Pythagoras of Samos. *Dictionary of Scientific Biography* 11:225.

3. A. C. Chibnall (1942). Amino-acid analysis and the structure of proteins. *Proc. Roy. Soc.* B131:136-160; J. A. Witkowski (1985). The magic of numbers. *TIBS* 10:139-141.

4. J. H. Jeans (1930). *The Mysterious Universe*, p. 144. New York: Macmillan.

5. A. D. Aleksandrov, A. N. Kolmogorov, and M. A. Lavrentiev (eds.) (1963). *Mathematics, its Content, Methods, and Meaning*. Cambridge, Mass.: MIT Press; C. B. Boyer (1968). *A History of Mathematics*. New York: Wiley; H.L.L. Busard (1997). Über die Entwicklung der Mathematik in Westeuropa zwischen 1100 und 1500. *NTM* NF5:211-235.

6. J. E. Hofmann (1971). Nicholas of Cusa. *Dictionary of Scientific Biography* 3:512-516. A. G. Debus (1968). Mathematics and nature in the chemical text of the Renaissance. *Ambix* 15:1-28.

7. M. Beretta (1993). The role of symbolism from alchemy to chemistry. In *Non-Verbal Communication in Science prior to 1900* (R. G. Mazzolini, ed.), pp. 279-319. Florence: Olschki.

8. A. Comte (1830-1842). *Cours de Philosophie Positive*. Paris: Bachelier. The quotation was taken from W. H. Brock (ed.) (1967). *The Atomic Debate*, p. 145, Leicester University Press, as given in the English translation of Comte's treatise by H. Martineau (1896), vol. 1, p. 289. See also B. Bensaude-Vincent (1999). Atomism and positivism. A legend about French chemistry. *Ann. Sci.* 56:81-94.

9. C. Bailey (1928). *The Greek Atomists and Epicurus*. Oxford University Press; K. Freeman (1949). *The Pre-Socratic Philosophers*. Oxford University Press.

10. A. E. Taylor (1960). *Plato: The Man and his Work*. 7th ed. London: Methuen; E. Hamilton and H. Cairns (eds.) (1961). *Plato: The Dialogues*. New York: Bollingen.

11. J. R. Partington (1948). The concepts of substance and chemical element. *Chymia* 1:109-121; J. E. Bolzan (1976). Chemical combination according to Aristotle. *Ambix* 23:134-144; S. M. Cohen (1996). *Aristotle on Nature and Incomplete Substance*. Cambridge University Press; J. van Brakel (1997). Chemistry as the science of the transformation of matter. *Synthese* 111:253-282.

12. G. Freudenthal (1995), *Aristotle's Theory of Material Substance: Heat and Pneuma, Form and Soul*. Oxford University Press; G.E.R. Lloyd (1996). *Aristotelian Explorations*. Cambridge University Press.

13. R. Boyle (1680). *The Sceptical Chymist etc.* 2nd ed. p. 354. Oxford: Hall. In his early writings, Boyle condemned the secrecy of the alchemists, but when he became one himself he followed their example; see

L. M. Principe (1992), Boyle's alchemical secrecy: codes, ciphers, and concealments. *Ambix* 39:63-74.

14. A. Lavoisier (1789). *Traité Elémentaire de Chimie.* p. xxiv - Paris: Cuchet. See T. L. Davis (1931), Boyle's conception of element compared with that of Lavoisier. *Isis* 16:82-91; F. A. Paneth (1962). The epistemological status of the chemical concept of the element. *Brit. J. Phil. Sci.* 13:1-14, 144-160.

15a. C. Bailey (1928). *The Greek Atomists and Epicurus.* Oxford University Press.

16. S. Sambursky (1959). *Physics of the Stoics.* London: Routledge & Kegan Paul; M. Lepidge (1988). The Stoic inheritance. In *A History of Twelfth-Century Western Philosophy* (P. Dronke, ed.), pp. 81-112. Cambridge University Press.

17. K. C. Bailey (1929, 1932). *The Elder Pliny's Chapters on Chemical Subjects.* London: Arnold; R. French and F. Greenaway (eds.) (1986). *Science in the Early Roman Empire: Pliny the Elder, his Sources and his Influence.* London: Crook Helm.

18. E. O. Lippmann (1919, 1931, 1954). *Entstehung und Ausbreitung der Alchemie.* Berlin: Springer (vol. 3, Weinheim: Verlag Chemie); L. Thorndike (1923-1958). *A History of Magic and Experimental Science.* New York: Columbia University Press; E. J. Holmyard (1957). *Alchemy.* Harmondsworth: Penguin Books; W. Eamon (1994). *Science and the Secrets of Nature.* Princeton University Press; R. Patai (1994). *The Jewish Alchemists.* Princeton University Press; J. B. Hershbell (1987). Democritus and the beginnings of Greek alchemy. *Ambix* 34:5-20.

19. A. J. Hopkins (1938). A study of the kerotakis process as given by Zosimos and later alchemical writers. *Isis* 29:326-354; M. Plessner (1976). Zosimos of Panapolis. *Dictionary of Scientific Biography* 14:631-633; C. Viano (1996). Aristote et l'alchimie grecque: La transmutation et le modèle aristotélicien entre théorie et pratique. *Rev. Hist. Sci.* 49:189-213; C. A. Wilson (1998). Pythagorean theory and Dionysian practice: The cultic and practical background to chemical experimentation in Hellenistic Egypt. *Ambix* 45:14-33.

20. A. I. Sabra (1987). The appropriation and subsequent naturalization of Greek science in medieval Islam. *Hist. Sci.* 25:223-243; P. Kraus (1986). *Jabir ibn Hayyan.* Paris: Belles Lettres; Y. Marquet (1988). *La Philosophie des Alchimistes et L'Alchimie des Philosophes.* Paris: Maisonneuve et Larose.

21. J. Ruska (1934). Die Alchemie des Avicenna. *Isis* 21:13-51.

22. G. Agricola (1912). *De Re Metallica* (translated by H. C. and L. H. Hoover). London: *The Mining Magazine*; H. Wilsdorf (1956). *Agricola und seine Zeit*. Berlin: VEB Verlag; M. Beretta (1997). Humanism and Chemistry: The Spread of Georgius Agricola's metallurgical writings. *Nuncius* 12:17-47; V. Biringuccio (1540). *De la Pirotechnia*. Venice: Rosinello; L. Ercker (1574). *Beschreibung Allerfürnemisten Mineralischen Ertzt und Bergwerke Arten*. Prague: Schwartz; A. Libavius (1597). *Alchemia*. Frankfurt: Kopffij. See R. P. Multhauf (1996). Operational practice and the emergence of modern concepts. *Science in Context* 9:241-249; C. Pliessner (1997). Basilius Valentinus und die Labortechnik um 1600. *Ber. Wissen.* 20:159-172.

23. A. C. Crombie (1953). *Robert Grosseteste and the Origins of Experimental Science*. Oxford: Clarendon Press; R. Brehm (1976). Roger Bacon's place in the history of alchemy. *Ambix* 23:53-58; D. C. Lindberg (1987). Science as handmaiden: Roger Bacon and the patristic tradition. *Isis* 74:509-530

24. S. Schaffer (1987). Godly men and mechanical philosophers: Souls and spirits in Restoration natural philosophy. *Osiris* 1:55-85; S. D. Snobelen (1999). Isaac Newton, heretic: the strategies of a Nicodemite. *Brit. J. Hist. Sci.* 32:381-419.

25. R. D. Gray (1952). *Goethe the Alchemist*. Cambridge University Press; G. Schwert (1998). Goethe als Chemiker. Berlin. Springer; T. H. Levere (1994). *Chemists and Chemistry in Nature and Society*. Aldershot: Variorum; S. R. Morgan (1990). Schelling and the origins of his *Naturphilosophie*. In *Romanticism and the Sciences* (A. Cunningham and N. Jardine, eds.), pp. 25-37. Cambridge University Press; C. G. Jung (1968). *Psychology and Alchemy*. 2nd ed. Princeton University Press; M. R. Newman (1996). *Decknamen* or pseudochemical language? Eirenaeus Philalethes and Carl Jung. *Rev. Hist.* 49:159-188.

26. J. R. Glauber (1659). *La Description des Fourneaux Philosophiques*. Paris: Jolly; J. Kunckel (1716). *Collegium Physico-Chemicum Experimentale*. Hambug: Heyl.

27. W. Pagel (1982). *Paracelsus*. 2nd ed. Basel: Karger; T. P. Sherlock (1948). The chemical work of Paracelsus. *Ambix* 3:33-63; R. P. Multhauf (1954), John of Rupescissa and the origin of medical chemistry. *Isis* 45:359-367; P. Meier (1993). *Paracelsus, Arzt und Prophet*. Zurich: Amman Verlag; D. von Engelhardt (1994). Paracelsus im

Urteil des 18. Jahrhunderts. *Gesnerus* 51:165-182; W. Pagel (1982). *Joan Baptista van Helmont.* Cambridge University Press; A. Clericuzio (1993). From van Helmont to Boyle: A study of the transmission of Helmontian chemical and medical theories in seventeenth-century England. *Brit. J. Hist. Sci.* 26:303-334.

28. A. G. Debus (1965). *The English Paracelsians.* London: Oldbourne; (1977). *The Chemical Philosophy. Paracelsian Science in the Sixteenth and Seventeenth Centuries.* New York: Science History Publications; (1991). *The French Paracelsians.* Cambridge University Press.

29. A. Perez-Ramos (1988). *Francis Bacon's Idea of Science and the Maker's Knowledge Tradition.* Oxford University Press. See M. Horton (1973). In defense of Bacon: A criticism of the critics of the inductive method. *Stud. Hist. Phil. Sci.* 4:241-278; P. Urbach (1987). *Francis Bacon's Philosophy of Science.* La Salle: Open Court; B. Vickers (1992). Francis Bacon and the progress of science. *J. Hist. Ideas* 53:495-517.

30. J. F. Scott (1952). *The Scientific Work of Descartes.* London: Taylor & Francis; A. C. Crombie et al. (1971). René du Perron Descartes. *Dictionary of Scientific Biography.* 4:51-65; A. Clericuzio (1990). A redefinition of Boyle's chemistry and corpuscular philosophy. *Ann. Sci.* 47:561-589; M. Hunter (ed.) (1994). *Robert Boyle Reconsidered.* Cambridge University Press; R. S. Westfall (1980). *Never at Rest: A Biography of Isaac Newton.* Cambridge University Press; B. J. Dobbs (1975). *The Foundations of Newton's Alchemy.* Cambridge University Press; (1991). *The Janus Face of Genius: The Role of Alchemy in Newton's Thought.* Cambridge University Press; J. Fauvel et al. (eds.) (1988). *Let Newton Be!* Oxford University Press; M. Hunter and S. Schaffer (eds.) (1989). *Robert Hooke: New Studies.* Woodbridge: Boydell Press; E. T. Drake (1996). *Restless Genius: Robert Hooke and his Earthly Thoughts.* Oxford University Press.

31. S. Shapin (1988). Robert Boyle and mathematics: reality, representation, and experimental science. *Sci. Context* 2:23-58; M. Oster (1992). The scholar and the craftsman revisited: Robert Boyle as aristocrat and artisan. *Ann. Sci.* 49:255-276.

32. Boyle (1674); quoted from A. C. Crombie (note 7), p. 965. See J. E. McGuire (1968). Force, active principles, and Newton's invisible realm. *Ambix* "Nature is a perpetual worker"; Newton's aether and eighteenth-century natural philosophy. *Ambix* 20:1-25; K. Hutchinson (1982). What happened to occult qualities in the Scientific Revolu-

tion? *Isis* 73:233-253; J. Henry (1986). Occult qualities and the experimental philosophy. *Hist. Sci.* 24:355-381; J. W. Wojcik (1997). *Robert Boyle and the Limits of Reason*. Cambridge University Press.

33. J. Mayow (1907). *Medico-Physical Works*. Edinburgh: Alembic Club. See A. Clericuzio (1995). The internal laboratory. The chemical reinterpretation of medical spirits in England. In *Alchemy and Chemistry in the 16th and 17th Centuries* (P. Rattansi and A. Clericuzio, eds.), pp. 52-83. Dordrecht: Kluwer.

34. T. S. Kuhn (1952). Robert Boyle and structural chemistry in the seventeenth century. *Isis* 43:12-36; M. Boas (1958). *Robert Boyle and Seventeenth-Century Chemistry*. Cambridge University Press.

35. B. J. Dobbs (1982). Newton's alchemy and his theory of matter. *Isis* 73:511-538; M. L. Gill and G. Lennox (eds.) (1994). *Self-Motion: From Aristotle to Newton*. Princeton University Press.

36. B. J. Dobbs (1975) (see note Ch.1:30); K. Figala (1977). Newton as an alchemist. *Hist. Sci.* 15:102-137; C. Webster (1982). *From Paracelsus to Newton: Magic and the Making of Modern Science*. Cambridge University Press; W. Bonsiepen (1993). Newtonian atomism and eighteenth-century chemistry. In *Hegel and Newtonanism* (M. J. Petry, ed.), pp. 595-608. Dordrecht: Kluwer.

37. T. Willis (1684). *Dr. Willis' Practice of Physic etc.* (translated by S. Portage). London: Dring, Harper & Leigh; H. Isler (1965), *Thomas Willis*. New York: Hafner; S. Hales (1733). *Statical Essays*. London: W. Innes et al.; H. Guerlac (1972). Stephen Hales. *Dictionary of Scientific Biography* 6:35-48; H. Boerhaave (1735). *Elements of Chemistry* (translated by T. Dallowe). London: Pemberton, Clarke, Miller & Gray; G. A. Lindeboom (1968). *Herman Boerhaave*. London: Methuen.

38. See Pagel (1982) (note Ch.1:27).

39. D. R. Oldroyd (1974). Some neo-Platonic and Stoic influences on mineralogy in the sixteenth and seventeenth centuries. *Ambix* 21:128-156.

40. R. P. Multhauf (1954). Medical chemistry and the "Paracelsians." *Bull. Hist. Med.* 28:101-126; W. Pagel (1956). Van Helmont's ideas on gastric digestion and the gastric acid. *Bull. Hist. Med.* 30:524-536; W. Schneider (1961). Der Wandel des Arzneischatzes im 17. Jahrhundert und Paracelsus. *Sudhoffs Arch.* 45:201-215.

41. A. G. Debus (1962). Solution analysis prior to Boyle. *Chymia.* 8:41-60; W. R. Newman (1994). Boyle's debt to corpuscular alchemy.

In *Robert Boyle Reconsidered* (M. Hunter, ed.), pp. 107-118. Cambridge University Press.

42. G. Freudenthal (1986). Die elektrische Anziehung im 17. Jahrhundert zwischen korpusculare und alchemischen Deutung. In *Die Alchemie in der europäischen Kultur etc.* (C. Meinel, ed.), pp. 315-326. Wiesbaden: Harrassowitz; (1990). The problem of cohesion between alchemy and natural philosophy: from unctuous moisture to phlogiston. In *Alchemy Revisited* (Z.R.W.M. van Martels, ed.), pp. 107-116. Leiden: Brill; L. M. Principe (1998). *The Aspiring Adept: Robert Boyle and his Alchemical Quest.* Princeton University Press; B. J. Dobbs (1971, 1973). Studies in the natural philosophy of Sir Kenelm Digby. *Ambix* 18:1-25; 20:143-163.

43. This experiment had been suggested a hundred years earlier by Nicholas of Cusa. See J. R. Partington (1961). *A History of Chemistry.* Vol. 2, pp. 223-224. London: Macmillan.

44. W. Pagel (1962). The wild spirit (Gas) of van Helmont and Paracelsus. *Ambix* 10:1-13; B. Joly (1996) L' alkahest, dissolvant universel ou quand la theorie rend possible une pratique impossible. *Rev. Hist. Sci.* 39:305-344.

45. W. R. Newman (1994). *Gehennical Fire: The Lives of George Starkey, an American Alchemist in the Scientific Revolution.* Cambridge, Mass.: Harvard University Press; A. Clericuzio (1996). Alchimie, philosophie corpusculaire, et minéralogique dans la *Metallographie* de John Webster. *Bull. Hist. Sci.* 49:287-304.

46. Webster (1671), pp. 83-84; quoted from Oldroyd (1974) (note 63), p. 151.

47. D. R. Oldroyd (1974). Mechanical mineralogy. *Ambix* 21:157-178. See also H. Metzger (1918). *La Genèse des Sciences des Cristaux.* Paris: Alcan.

48. G. Scherz (1976). Niels Stensen. *Dictionary of Scientific Biography* 13:30-35.

49. H. Seifert (1954). Nicolaus Steno als Bahnbrecher der modernen Kristallographie. *Sudhoffs Arch.* 38:29-47.

50. B. Rochot (1972). Pierre Gassendi. *Dictionary of Scientific Biography* 5:284-290.

51. H. Metzger (1918). *La Genèse de la Science des Cristaux.* Paris: Alcan; J. G. Burke (1966). *Origins of the Science of Crystals.* Berkeley: University of California Press; J. J. Burckhardt (1988). *Die Symmetrie*

der Kristalle von Renè Just Haüy zur kristallographischen Schule in Zürich.
Basel: Birkhäuser.

52. L. M. Principe (note Ch.1:42); W. R. Newman (1999). Alchemical symbolism and concealment: the chemical house of Libavius. In *The Architecture of Science* (P. Galison and E. Thompson, eds.), pp. 59-77. Cambridge, Mass.: MIT Press.

Chapter Two

1. J. Eklund (1975). *The Incompleat Chymist*. Washington, D.C.: Smithsonian Institution. See T. S. Patterson (1937). Jean Beguin and his "Tyrocinium Chemicum." *Ann. Sci.* 2:243-298.

2. M. P. Crosland (1962). *Historical Studies in the Language of Chemistry*. London: Heinemann; O. Hannaway (1975). *The Chemists and the Word*. Baltimore: Johns Hopkins University Press.

3. H. Schelenz (1911). *Geschichte der pharmazeutisch-chemischen Destilliergeräte*. Berlin: Springer; R. J. Forbes (1948). *Short History of the Art of Distillation*. Leiden: Brill; A. J. Liebmann (1956). History of distillation. *J. Chem. Educ.* 33:166-173; R. Multhauf (1956). Significance of distillation in Renaissance medical chemistry. *Bull. Hist. Med Chem.* 41:1321-1324.

4. W. Eamon (1980). New light on Robert Boyle and the discovery of color indicators. *Ambix* 27:204-209; A. A. Baker (1964). A history of indicators. *Chymia* 9:147-167.

5. R. P. Multhauf (1962). The use of the balance in chemistry. *Proc. Am. Phil. Soc.* 106:210-227; R. Halleux (1986). L'alchimiste et l'essayeur. In *Die Alchemie in der europäischen Kultur- und Wissenschaftsgeschichte* (C. Meinel, ed.), pp. 277-291. Wiesbaden: Harrasowitz; M. Crosland (1996). Changes in chemical concepts and language in the seventeenth century. *Sci. Context* 9:222-240.

6. F. Szabadvary (1966). *Geschichte der Analytichen Chemie*. Budapest: Akademiai Kiado.

7. Stahl quoted from D. R. Oldroyd (1973). An examination of G. E. Stahl's Philosophical Principles of Universal Chemistry. *Ambix* 36-52 (39); Rouelle quoted from Eklund (1975) (note Ch.2:1), p. 2.

8. F. L. Holmes (1989). *Eighteenth-Century Chemistry as an Investigative Enterprise*. Berkeley: Office of History of Science and Technology; U. Klein (1996). The chemical workshop tradition and the experimental practice: *Sci. Context* 9:251-287.

9. H. Metzger (1926). La philosophie de la matière chez Stahl et ses disciples. *Isis* 8:427-464; (1927). La thèorie de la composition des sels et la thèorie de la combustion d'après Stahl et ses disciples. *Isis* 9:294-325; (1930). *Newton, Stahl, Boerhaave et la Doctrine Chimique.* Paris: Alcan.

10. J. R. Partington and D. McKie (1937-1939). Historical studies on the phlogiston theory. *Ann. Sci.* 2:361-404; 3:1-58, 337-371; 4:113-149. W. H. Suddoth (1978). Eighteenth-century identification of electricity with phlogiston. *Ambix* 25:131-147.

11. Quoted from R. Siegfried and B. J. Dobbs (1968). Composition, a neglected aspect of the chemical revolution. *Ann. Sci.* 24:275-293 (276). See L.J.M. Coleby (1938). *The Chemical work of P. J. Macquer.* London: Allen and Unwin; R. G. Neville (1966). Macquer and the first chemical dictionary, 1766. *J. Chem. Educ.* 43:486-490.

12. J. Parascandola and A. J. Ihde (1969). History of the pneumatic trough. *Isis* 60:351-361.

13. J. R. Partington (1962). *A History of Chemistry,* vol. 3, p. 123. London: Macmillan.

14. M. Kerker (1955). Herman Boerhaave and the development of pneumatic chemistry. *Isis* 46:36-49.

15. Metzger (1930) (note Ch.2:9); J. C. Gregory (1934). *Combustion from Heraclitus to Lavoisier.* London: Arnold; R. Love (1972). Some sources of Herman Boerhaave's concept of fire. *Ambix* 19:157-174.

16. J. Black (1944). *Experiments on Magnesia Alba etc.* Edinburgh: Alembic Club; L. Dobbin (1935). Joseph Black's inaugural dissertation. *J. Chem. Educ.* 12:268-273; W. Ramsay (1918). *The Life and Letters of Joseph Black.* London: Constable; H. Guerlac (1957). Joseph Black and fixed air: a bicentennial retrospective. *Isis* 48:433-456; (1970). Joseph Black. *Dictionary of Scientific Biography* 2:173-183; A.D. C. Simpson (ed.) (1980). *Joseph Black 1728-1799: A Commemorative Symposium.* Edinburgh: Royal Scottish Museum; J. R. Partington (1960). Joseph Black's "Lectures on the Elements of Chemistry." *Chymia* 6:27-67.

17. J. Ruska (1928). Der Salmiak in der Geschichte der Alchemie. *Z. angew. Chem.* 41:1321-1324; R. P. Multhauf (1965). Sal ammoniac, a case history in industrialization. *Technol. Cult.* 6:569-586. See R. Koch (1924). Glauber. In *Das Buch der Grossen Chemiker* (G. Bugge, ed.), vol. 1, pp. 151-172. Berlin: Verlag Chemie.

18. J. R. Partington (1962). *A History of Chemistry.* Vol. 2, pp. 637-652. London: Macmillan; See B. Bernadette-Vincent (1998). *Eloge du Mixte.* Paris: Hachette.

19. Quoted from Oldroyd (1973) (note Ch.2:7), p. 43.

20. H. Guerlac (1959). Some French antecedents of the Chemical Revolution. *Chymia* 5:73-112; M. Fichman (1971). French Stahlism and chemical studies on air. *Ambix* 18:94-112; J. B. Gough (1988). Lavoisier and the fulfillment of the Stahlian revolution. *Osiris* [2] 4:15-33; J. Simon (1998). The Chemical Revolution and pharmacy. *Ambix* 45:1-13.

21. J. P. Constant (1952). *L'Enseignment de la Chimie au Jardin Royal des Plantes de Paris*. Cahours: Coueslant.

22. H. Metzger (1923). *Les Doctrines Chimiques en France du XVIIe à Fin du XVIIIe Siècle*. Paris: Presses Universitaires; J. C. Powers (1998). 'Ars sine arte:' Nicholas Lemery and the end of alchemy in eighteenth-century France. *Ambix* 45:163-189; M. Bougard (1999). *La Chimie de Nicolas Lemery*. Turnhout: Brefols.

23. J. R. Partington (1962). *A History of Chemistry*, vol. 3, pp. 28-41; A. M. Duncan (1970), The functions of affinity tables and Lavoisier's list of elements. *Ambix* 17:28-42; (1996). *Laws and Order in Eighteenth-Century Chemistry*. Oxford: Clarendon Press; U. Klein (1994). Origin of the concept of the chemical compound. *Sci. Context* 7:163-204; (1995). E. F. Geoffroy's table of different "rapports observed between different chemical substances = a reinterpretation. *Ambix* 42:79-100; F. L. Holmes (1996). The communal context for Etienne-François Geoffroy's "Table des Rapports." *Sci. Context*. 9:289-311.

24. J. A. Schufle (1967). Torbern Bergman, earth chemist. *Chymia* 12:59-97; W. A. Smeaton (1963). Guyton de Morveau and chemical affinity. *Ambix* 11:55-64.

25. R. Rappaport (1960). G. F. Rouelle: An eighteenth-century chemist and teacher. *Chymia* 6:68-101; (1961). Rouelle and Stahl—The phlogistic revolution in France. *Chymia* 7:73-102.

26. Siegfried and Dobbs (1968) (note Ch.2:11).

27. A. Lavoisier (1789). *Traité Elémentaire de Chimie*, vol. 1, p. 38. Paris: Cuchet. The dates Lavoisier cited for his research articles were those of submission, not of publication (frequently two or three years later).

28. C. W. Scheele (1980). *Chemical Observations and Experiments on Air and Fire* (translated by J. R. Forster). London: Johnson; (1931). *The Collected Papers of Carl Wilhelm Scheele* (translated by L. Dobbin). London: Bell.

29. U. Bokland (1975). Carl Wilhelm Scheele. *Dictionary of Scien-*

tific Biography 12:143-150; W. A. Smeaton (1970). Torbern Olof Bergman. Ibid., 2:4-8.

30. S. J. French (1950). The Chemical Revolution—the second phase. *J. Chem. Educ.* 27:83-89; R. Siegfried (1988). The Chemical Revolution in the history of chemistry. *Osiris* [2] 4:34-50; J. G. McEvoy (1988). Continuity and discontinuity in the Chemical Revolution. *Osiris* [2] 4:195-213.

31. R.E.W. Maddison and F. R. Maddison (1956). Joseph Priestley and the Birmingham riots. *Notes Roy. Soc.* 12:98-136; R. B. Rose (1960). The Priestley riots of 1791. *Past and Present* 18:69-88. J. G. McEvoy and J. E. McGuire (1975). God and nature: Priestley's way of rational dissent. *Hist. Stud. Phys. Sci.* 6:325-404.

32. J. Priestley (1774). *Experiments and Observations on Different Kinds of Air.* London: Lohnson; (1777). An account of further discoveries on air. *Phil. Trans.* 65:384-394; J. G. McEvoy (1978, 1979). Joseph Priestley, "Aerial Philosopher." *Ambix* 25:1-55, 93-116, 153-175; 26:16-38; R. E. Schofield (1997). *The Enlightenment of Joseph Priestley: A Study of his Life and Work from 1733 to 1773.* University Park, Pa.: Pennsylvania State University Press.

33. J. Ingen-Housz (1780). *Expériences sur les Végétaux.* Paris: Didot; J. Senebier (1788). *Expériences sur l'Action de la Lumière Solaire dans la Végétation.* Geneva: Barde, Marget; N. T. de Saussure (1804). *Recherches Chimiques sur la Végétation.* Paris: Nyon.

34. T. L. Davis (1927). The last stand of phlogiston: Priestley's defense of the doctrine after his removal to America. In *Studien zur Geschichte der Chemie* (J. Ruska, ed.), pp. 132-147. Berlin: Springer; S. Soloveichik (1962). The last fight for phlogiston and the death of Priestley. *J. Chem. Educ.* 39:644-646; M. E. Conlin (1996). Joseph Priestley's American defense of phlogiston reconsidered. *Ambix* 43:129-145.

35. J. R. Partington (note Ch.2:13), pp. 268-271, 284-286.

36. R. McCormmach (1961). Henry Cavendish. A study of rational empiricism in eighteenth-century natural philosophy. *Isis* 60:293-306.

37. G. Wilson (1851). *The Life of the Honourable Henry Cavendish.* London: Cavendish Society; A. J. Berry (1960). *Cavendish, his Life and Scientific Work.* London: Hutchinson; C. Jungnickel and R. McCormmach (1996). *Cavendish.* Philadelphia: American Philosophical Society.

38. H. Cavendish (1926). *Experiments on Air.* Edinburgh: Alembic Club. Similar experiments were performed in 1784 by Gaspard Monge.

For the eudiometer, see R. Bud and D. J. Warner (eds.) (1998). *Instruments of Science*, pp. 232-234. New York: Garland.

39. A. Lavoisier (1862). *Oeuvres de Lavoisier.* pp. 334-359. Paris: Imprimerie Impériale.

40. H. Cavendish (1926) (note Ch.2:37), pp. 35-36.

41. Ibid., p. 38.

42. H. Kopp (1843-1847). *Geschichte der Chemie.* pp. 264-271. Braunschweig: Vieweg; H. Edelstein (1948). Priestley settles the water controversy. *Chymia* 1:123-137; M. Dauman and D. Duveen (1959), Lavoisier's relatively unknown large-scale decomposition and synthesis of water, February 27 and 28, 1785. *Chymia* 5:113-129; R. E. Siegfried (1961). James Watt's letter to Joseph Priestley 26 April 1783. *Ambix* 9:294-300.

43. R. Fox (1971). *The Caloric Theory of Heat from Lavoisier to Regnault.* Oxford University Press; R. McCormmach (1988). Henry Cavendish on the theory of heat. *Isis* 79:37-67; W.E.K. Middleton (1966). *A History of the Thermometer.* Baltimore: Johns Hopkins University Press; T. H. Lodwig and W. A. Smeaton (1974). The ice calorimeter of Lavoisier and Laplace and some of its critics. *Ann. Sci.* 31:1-18.

Chapter Three

1. My translation of French quotation in J. B. Gough (1982). Some early references to revolutions in chemistry. *Ambix* 106-109 (109). A. Levin (1984). Venel, Lavoisier, Fourcroy, Cabanis and the idea of scientific revolution. *Hist. Sci.* 22:303-320.

2. Mme. Lavoisier dominates the famous portrait of the pair by David. It now hangs in the Metropolitan Museum of Art in New York. I had frequent opportunity to study it at the library of the Rockefeller Institute for Medical Research during 1934-1945.

3. Of the many general biographies, apart from that of E. Grimaux (1888). *Lavoisier 1743-1794.* Paris: Alcan, the more recent ones are those by D. McKie (1952). *Antoine Lavoisier.* London: Constable; M. Daumas (1955). *Lavoisier, Théoricien et Expérimenteur.* Paris: Presses Universitaires; H. Guerlac (1973). Antoine Laurent Lavoisier. *Dictionary of Scientific Biography* 8:66-91; B. Bensaude-Vincent (1993). *Lavoisier: Mémoires d'une Revolution.* Paris: Flammarion; A. Donovan (1994). *Antoine Lavoisier.* Oxford: Blackwell. The one by J. P. Poirier

(1993). *Antoine Laurent de Lavoisier*. Paris: Pygmalion has an excellent account of Lavoisier's services to the state.

4. A. N. Meldrum (1933). Lavoisier's early work in science. *Isis* 19:330-363; R. Rappaport (1968). Lavoisier's geologic activities 1763-1792. *Isis* 58:375-384; J. B. Gough (1983). Lavoisier's memoirs on the nature of water and their place in the Chemical Revolution. *Ambix* 30:89-106.

5. C. A. Lopez (1960). Saltpetre, tin, and gunpowder: Addenda to the correspondence of Lavoisier and Franklin. *Ann. Sci.* 16:83-94; P. Bret (1996). Lavoisier et l'apport de la chimie académique à l'industrie des poudres et salpêtres. *Arch. Int. Hist. Sci.* 46:57-74; V. Viel (1996). Le salon et le laboratoire de Lavoisier à l'Arsenal, etc. *Hist. Sci. Med.* 30:32-34.

6. A. N. Meldrum (1930). *The Eighteenth-Century Revolution in Science—The First Phase*. London: Longmans, Green; H. Guerlac (1961). *Lavoisier—The Crucial Year*. Ithaca: Cornell University Press; F. L. Holmes (1997). *Antoine Lavoisier—The Next Crucial Year*. Princeton University Press. See P. Bret (1995). Trois décennies d'études lavoisiennes. *Rev. Hist. Sci.* 48:169-197.

7. M. Daumas (1950). Les appareils d'experimentation de Lavoisier. *Chymia* 3:45-62; (1963). Precision of measurement and physical and chemical research in the eighteenth century. In *Scientific Change* (A. C. Crombie, ed.), pp. 418-430. London: Heinemann; R. E. Oesper (1940) Some famous balances. *J. Chem. Educ.*17:312-323; R. Bud and D. J. Warner (1998) (eds.). Chemical balances. In *Instruments of Science*, pp. 45-47. New York: Garland.

8. C. C. Gillispie (1960). *The Edge of Objectivity*, pp. 202-259. Princeton University Press; H. Guerlac (1961). Quantification in chemistry. *Isis* 52:194-214; A. Lundgren (1975). The changing role of numbers in 18th-century chemistry. In *The Quantifying Spirit in the 18th Century* (T. Frangsmyr et al., eds), pp. 245-266. Berkeley: University of California Press; T. H. Levere (1992). *Chemists and Chemistry in Nature and Society 1770-1878*, pp. 313-332. Aldershot: Variorum; F. L. Holmes (1997) (note Ch.3:6).

9. I. Freund (1904). *The Study of Chemical Composition*, p. 60. Cambridge University Press.

10. N. R. Hanson (1965). Galileo's discoveries in dynamics. *Science* 147:471-478 (474); A. Koyré (1968). *Metaphysics and Measurement*, p. 94. Cambridge, Mass. Harvard University Press.

11. A. Lavoisier (1789) (note Ch.2:27), pp. 140-141. See F. L. Holmes (1985). *Lavoisier and the Chemistry of Life*, pp. 326-352. Madison: University of Wisconsin Press.

12. A. Lavoisier (1789) (note Ch.2:27), p. 151.

13. A. Harden (1923). *Alcoholic Fermentation*, 3rd ed., p. 3. London: Longmans, Green.

14. F. L. Holmes (1997) (note Ch.3:6), p. 89.

15. One of Pasteur's objections to the Lavoisier equation was that succinic acid is a by-product in alcoholic fermentation; see L. Pasteur (1860). Mémoire sur la fermentation alcoolique. *Ann. Chim.* [3] 58:323-426.

16. C. E. Perrin (1990). Chemistry as a par of physics. *Isis* 81: 259-270; F. L. Holmes (1995). The boundaries of Lavoisier's knowledge. *Rev. Hist. Sci.* 48:9-48.

17. H. Guerlac (1961) (note Ch.3:6); C. E. Perrin (1986). Lavoisier's thoughts on calcination and combustion, 1772-1773. *Isis* 77:647-666; (1989). Document, text, and myth: Lavoisier's crucial year revisited. *Brit. J. Hist. Sci.* 22:3-25; F. L. Holmes (1988). Lavoisier's conceptual passage. *Osiris* [2] 4:82-92.

18. C. E. Perrin (1987). Revolution or reform? The Chemical Revolution and eighteenth century concepts of scientific change. *Hist. Sci.* 25:395-423; (1988). Research traditions, Lavoisier and the Chemical Revolution. *Osiris* [2] 4:53-81; F. L. Holmes (1997). What was the Chemical Revolution about? *Bull. Hist. Chem.* 20:1-9.

19. This frequently cited sentence appears to be a later revision of the one in the original document. J. R. Partington (note Ch.2:13), p. 385; H. Guerlac (1961). A curious Lavoisier episode. *Chymia* 7:103-108.

20. F. L. Holmes (1997) (note Ch.3:6), p. 139.

21. E.W.J. Neave (1951). Chemistry in Rozier's Journal. III. Pierre Bayen. *Ann. Sci.* 7:144-148; J. R. Partington (1962) (note Ch.2:13), pp. 396-399; E. McDonald (1970). Pierre Bayen. *Dictionary of Scientific Biography* 1:529-530.

22. S. J. French (1950 (note Ch.2:30); H. Guerlac (1957). Joseph Priestley's first papers and their reception in France. *J. Hist. Med.* 12:1-12; S. S. Toulmin (1957). Crucial experiments: Priestley and Lavoisier. *J. Hist. Ideas* 18:205-220.

23. C. A. Culotta (1972). Respiration and the Lavoisier tradition:

Theory and modification. *Trans. Am. Phil. Soc.* NS 62, part 3, pp. 1-41. F. L. Holmes (1985). *Lavoisier and the Chemistry of Life.* Madison, Wis.: University of Wisconsin Press.

24. J. R. Partington (note Ch.2:13), p. 414; M. Crosland (1973). Lavoisier's theory of acidity. *Isis* 64:306-325; H. E. Le Grand (1974). Ideas on the composition of muriatic acid and their relevance to the oxygen theory of acidity. *Ann. Sci.* 31:213-225.

25. R. Fox (1971) (note Ch.2:43); R. J. Morris (1972). Lavoisier and the caloric theory. *Brit. J. Hist. Sci.* 6:1-38.

26. J. R. Partington and D. McKie (1937-1939) (note Ch.2:10); D. Allchin (1992). Phlogiston after oxygen. *Ambix* 39:110-116.

27. M. Friedman (1992) (note 21) Foreword, pp. 264-341.

28. B. Bensaude-Vincent (1990). A view of the Chemical Revolution through contemporary text books—Lavoisier, Fourcroy, and Chaptal. *Brit. J. Hist. Sci.* 23:435-460.

29. L.B.G. de Morveau, A. Lavoisier, C. L. Berthollet, and A. F. Fourcroy (1787). *Méthode de Nomenclature Chimique.* Paris: Cuchet; W. A. Smeaton (1954). The contributions of P. J. Macquer, T. O. Bergman, and L. B. Guyton de Morveau to the reform of the chemical nomenclature. *Ann. Sci.* 10:87-106; J. W. Llana (1985). A contribution of natural history to the Chemical Revolution in France. *Ambix* 32:71-91; J. Riskin (1998). Rival idioms for a revolutionized science and a republican citizenry. *Isis* 89:203-232. The four authors of the *Méthode* formed the nucleus of the editorial board of the *Annales de Chimie*; see S. Court (1972). The Annales de Chemie, 1789-1815. *Ambix* 19:113-128; M. Crosland (1994). *In the Shadow of Lavoisier: The Annales de Chimie and the Establishment of a New Science.* London: British Society for the History of Science.

30. A. Lavoisier (1789). *Traité Élémentaire de Chimie.* p. 189. Paris: Cuchet; C. E. Perrin (1973), Lavoisier's table of simple substances. *Ambix* 20:95-105; R. Siegfried (1982). Lavoisier's table of simple substances: its origin and interpretation. *Ambix* 29:29-48.

31. A. Lavoisier (1789) (note Ch.3:30), p. xxix.

32. M. Beretta (1993). *The Enlightenment of Matter*, pp. 245-322. Canton, Mass.: Science History Publications; S. Lindroth (1973). Carl Linneaus. *Dictionary of Scientific Biography* 8:374-381.

33. A. Lavoisier (1789) (note Ch.3:30), pp. xiii-xiv.

34. A. Lavoisier (1782). Mémoire sur l'affinité du Principe oxygène.

Mém. Acad. Roy. Sci., p. 530. Quotation from A. M. Duncan (1970). The functions of affinity tables and Lavoisier's list of elements. *Ambix* 17:28-42 (29).

35. A. Lavoisier (1789) (note Ch.3:30), pp. 132-138.

36. H. Kopp (1869). *Beiträge zur Geschichte der Chemie*, pp. 141-144. Braunschweig: Vieweg.

37. J. R. Partington (note Ch.2:13), pp. 462-466; I. B. Cohen (1976). The eighteenth-century origins of the concept of scientific revolution. *J. Hist. Ideas* 37:257-288; H. Guerlac (1976). The Chemical Revolution: A word from Monsieur Fourcroy. *Ambix* 23:1-4; C. A. Perrin (1981). The triumph of the "anti-phlogistians." In *The Analytic Spirit* (H. Woolf, ed.), pp. 40-63. Ithaca: Cornell University Press; (1982). A reluctant catalyst: Joseph Black and the Edinburgh reception of Lavoisier's chemistry. *Ambix* 29:141-176.

38. H. Guerlac (1976). Chemistry as a branch of physics: Laplace's collaboration with Lavoisier. *Hist. Stud. Phys. Sci.* 7:197-223; A. Donovan (1988). Lavoisier and the origins of modern chemistry. *Osiris* [2] 4:214-231; E. M. Melhado (1989). Toward an understanding of the Chemical Revolution. *Knowledge and Society* 8:123-137; C. E. Perrin (1988). The Chemical Revolution—Shifts in guiding assumptions. In *Scrutinizing Science*. (A. Donovan et al., eds.), pp. 105-124. Dordrecht: Kluwer; (1988) (note 135).

39. A. Wurtz (1869). *Histoire des Doctrines Chimiques depuis Lavoisier jusqu'à nos Jours*. Paris: Hachette; M. Berthelot (1890). *La Révolution Chimique—Lavoisier*. Paris: Alcan; B. Bensaude-Vincent (1996). Between history and memory: Centennial and bicentennial images of Lavoisier. *Isis* 87:481-499.

Chapter Four

1. J. R. Partington (1951, 1953). Jeremias Benjamin Richter and the law of reciprocal proportions. *Ann. Sci.* 7:173-198; 9:289-317.

2. J. B. Richter (1792). *Anfangsgründe der Stöchymetrie oder Messkunst chemischer Elemente*. Breslau: Korn. Quotation from J. R. Partington (1951) (note Ch.4:1), pp. 179-180.

3. J. R. Partington (note Ch.2:13), pp. 672-673; J. J. Berzelius (1819). *Essai sur la Théorie des Proportions Chimiques etc.* Paris: Méquignon-Marvis; M. Daumas (1946). *L'Acte Chimique*. Brussels: Sablon;

M. W. Lindauer (1962). The evolution of the concept of chemical equilibrium from 1775 to 1929. *J. Chem. Educ.* 39:384-390.

4. B. W. Keyser (1990). Between science and craft: The case of Berthollet and dyeing. *Ann. Sci.* 47:213-260.

5. M. Sadoun-Goupil (1977). *Le Chimiste Claude-Louis Berthollet.* Paris: Vrin; H. E. Le Grand (1976). Berthollet's Essai de Statique Chimique and acidity. *Isis* 67:229-238; P. Grapí and M. Izquierdo (1997). Berthollet's conception of chemical change in context. *Ambix* 44:113-130; M. P. Crosland (1967). *The Society of Arcueil: A View of French Science at the Time of Napoleon I.* London: Heinemann.

6. J. R. Partington (1959). Berthollet and the antiphlogistic theory. *Chymia* 5:130-137: H. E. Le Grand (1975). The "conversion" of C. L. Berthollet to Lavoisier's chemistry. *Ambix* 22:58-70.

7. J. R. Partington (1951) (note Ch.4:1), p. 188.

8. I. Freund (note Ch.3:9), pp. 109-151; F. L. Holmes (1962). From elective affinities to chemical equilibrium. *Chymia* 8:105-145; S. C. Kapoor (1965). Berthollet, Proust, and proportions. *Chymia* 10:53-110.

9. J. R. Partington (note Ch.2:13), pp. 755-622; F. Greenaway (1966). *John Dalton and the Atom.* Ithaca: Cornell University Press; D.S.L. Cardwell (1968). *John Dalton and the Progress of Science.* Manchester University Press; E. C. Patterson (1970). *John Dalton and the Atomic Theory.* Garden City, N.Y.: Doubleday; A. Thackray (1972). *John Dalton.* Cambridge, Mass.: Harvard University Press. The Dalton manuscript collection in Manchester was destroyed during a German air raid in December 1940.

10. Both quotations taken from J. R. Partington (note Ch.2:13), p. 758.

11. H. E. Roscoe and A. Harden (1896). *A New View of the Origin of Dalton's Atomic Theory.* London: Macmillan; L. K. Nash (1956). The origin of Dalton's chemical atomic theory. *Isis* 47:101-116; H. Guerlac (1961). Some Daltonian doubts. *Isis* 52:544-554; A. Thackray (1966). The origin of Dalton's chemical atomic theory: Daltonian doubts resolved. *Isis* 57:35-55; The emergence of Dalton's chemical atomic theory 1801-1808. *Brit. J. Hist. Sci.* 3:1-22, (1970). *Atoms and Powers.* Oxford University Press; S. H. Mauskopf (1970). Haüy's model of chemical equivalence: Daltonian doubts exhumed. *Ambix* 17:182-191; R. S. Fleming (1974). Newton, gases, and Daltonian chemistry: the foundations of combination in definite proportions. *Ann. Sci.* 31:561-574; T. Cole,

Jr. (1978). Dalton, mixed gases, and the origin of the chemical atomic theory. *Ambix* 25:117-130; A. J. Rocke (1984). *Chemical Atomism in the Nineteenth Century: From Dalton to Cannizzaro*. Columbus: Ohio State University Press; K. Fujii (1986). The Berthollet-Proust controversy and Dalton's chemical theory 1800-1820. *Brit. J. Hist. Sci.* 19:177-200.

12. Quotation from H. E. Roscoe and A. Harden (note Ch.4:11), p. 13.

13. M. P. Crosland (1972). Joseph Louis Gay-Lussac. *Dictionary of Scientific Biography* 5:317-327.

14. J. Dalton (1808). *A New System of Chemical Philosophy*. London: Bickerstaff; J. Dalton, W. H. Wollaston, and T. Thomson (1923) *Foundations of the Atomic Theory*; J. Dalton, J. L. Gay-Lussac, and A. Avogadro (1950). *Foundations of the Molecular Theory*. Edinburgh: Alembic Club.

15. T. S. Wheeler and J. R. Partington (1963). *The Life and Work of William Higgins*. London: Pergamon.

16. M. E. Weeks (1968). *Discovery of the Elements*. 7th ed. Easton, Pa.: Journal of Chemical Education.

17. A. Treneer (1963). *The Mercurial Chemist: A Life of Humphry Davy*. London: Methuen; D. Knight (1992). *Humphry Davy: Science and Power*. Oxford: Blackwell. J. Z. Fullmer (2000). *Young Humphry Davy*. Philadelphia: American Philosophical Society.

18. T. H. Levere (1994). *Chemists and Chemistry in Nature and Society 1770-1878*. pp. 349-378. Aldershot: Variorum.

19. D.S.L. Cardwell (1972). *The Organization of Science in England*. 2nd ed. London: Heinemann; J. Golinski (1992). *Science as Public Culture*. Cambridge University Press.

20. C. A. Russell (1959, 1963). The electrochemical theory of Humphry Davy. *Ann. Sci.* 15:1-25; 19:255-271.

21. H. Davy (1935). *The Decomposition of the Fixed Alkalies and Alkaline Earths*, p. 43. Edinburgh: Alembic Club; R. Siegfried (1963). The discovery of potassium and sodium, and the problem of the chemical elements. *Isis* 54:247-258.

22. H. Davy (1929). *The Elementary Nature of Chlorine*. Edinburgh: Alembic Club.

23. Quoted from R. Siegfried (1964). The phlogistic conjectures of Humphry Davy. *Chymia* 9:117-124 (119-120).

24. R. Siegfried (1959). The chemical philosophy of Humphry Davy. *Chymia* 5:193-201.

25. W. Prout, J. S. Stas, and C. Marignac (1947). *Prout's Hypothesis.* Edinburgh: Alembic Club; R. Siegfried (1956). The chemical basis for Prout's hypothesis. *T. Chem. Educ.* 33:263-266.

26. W. H. Brock (1965). The life and work of William Prout. *Med. Hist.* 9:101-126; (1975). William Prout. *Dictionary of Scientific Biography* 11:172-174.

27. W. Prout (1947) (note Ch.4:25), p. 37.

28. Ibid., pp. 45, 58. D. C. Goodman (1969). Wollaston and the atomic theory of Dalton. *Hist. Stud. Phys. Sci.* 1:37-59; (1976). William Hyde Wollaston. *Dictionary of Scientific Biography,* 14:486-494. For Thomson, see I. Freund (1904) (note Ch.3:9), pp. 596-598.

29. W. Brock (1975) (note Ch.4:26), p. 173.

30. P. Collins (1975). Humphry Davy and heterogeneous catalysis. *Ambix* 22:205-217; (1976). Johann Wolfgang Döbereiner and heterogeneous catalysis. *Ambix* 23:96-115. See also J. Stradins (1964). The work of Theodore Grotthus and the invention of Davy's safety lamp. *Chymia* 9:125-145.

31. N. W. Fisher (1982). Avogadro, the chemists, and historians of chemistry. *Hist. Sci.* 20:77-102, 212-231; M. Morselli (1984). *Amedeo Avogadro.* Dordrecht: Kluwer; J. R. Hofmann (1995). *André-Marie Ampère.* Oxford: Blackwell; M. Scheidecker-Chevalier (1997); L'hypothèse d'Avogadro (1811) et d'Ampère (1814): la distinction entre atome/molécule et la théorie de la combinaison chimique. *Rev. Hist. Sci.* 50:159-194. See also I. Freund (1904) (note Ch.3:9), pp. 301-329.

32. For a correction, see H. Kopp (1873). *Die Entwicklung der Chemie in der neueren Zeit,* pp. 349-352, 838-839. Munich: Oldenbourg.

Chapter Five

1. W. A. Smeaton (1962). *Fourcroy, Chemist and Revolutionary.* Cambridge: Heffer; G. Kersaint (1966). *Antoine François Fourcroy, sa Vie et son Oeuvre.* Paris: Editions du Muséum; A. B. Costa (1962). *Michel Eugène Chevreul, Pioneer of Organic Chemistry.* Madison, Wis.: University of Wisconsin Press.

2. D. C. Goodman (1972). Chemistry and the two organic kingdoms of nature in the nineteenth century. *Med. Hist.* 16:113-130.

3. P. Walden (1927). Von der Iatrochemie zur "organischen" Chemie. *Z. angew. Chem.* 40:1-16; E. O. von Lippmann (1934).

Alter und Herkunft des Namens "Organische Chemie." *Chem. Z.* 58:1009-1011, 1031-1032.

4. J. J. Berzelius (1813). *A View of the Progress and Present State of Animal Chemistry* (translated by G. Brunnmark). London: Jatchard, Johnson, and Boosey; A. J. Rocke (1992). Berzelius's animal chemistry: from physiology to organic chemistry. In *Enlightenment Science in the Romantic Era* (E. M. Melhado and T. Frängsmyr, eds.). pp. 107-131. Cambridge University Press.

5. J. E. Jorpes (1966). *Jac. Berzelius: His Life and Work*, pp. 65-66. Stockhom: Almquist & Wiksell.

6. C. A. Russell (1968). Berzelius and the development of the atomic theory. In *John Dalton and the Progress of Science* (D.S.L. Cardwell, ed.), pp. 270-278. Manchester University Press; H. M. Leicester (1970). Jöns Jacob Berzelius. *Dictionary of Scientific Biography* 2:90-97; E. M. Melhado (1981). *Jacob Berzelius: The Emergence of his Chemical System.* Madison, Wis.: University of Wisconsin Press; E. M. Melhado and T. Frängsmyr, eds.) (1992) (note Ch.5:4).

7. J. J. Berzelius (1819). *Essai sur la Théorie des Proportions Chimiques et sur l'Influence Chimique de l'Elecricité.* Paris: Méquignon-Marvis.

8. M. P. Crosland (1962), *Historical Studies in the Language of Chemistry*, pp. 270-281. London: Heinemann.

9. See the table in J. R. Partington (1964). *A History of Chemistry.* Vol. 4, p. 166. London: Macmillan.

10. R. K. Fitzgerel and F. H. Verhoek (1960). The law of Dulong and Petit. *J. Chem. Educ.* 37:535-549. See also I. Freund (note Ch.3:9), pp. 361-384; J. R. Partington (note Ch.5:9), 160-166.

11. H. W. Schütt (1974). Zum Prioritätsproblem der Entstehung des Isomorphismus. *Physis* 16:5-22; E. M. Melhado (1980). Mitscherlich's discovery of isomorphism. *Hist. Stud. Phys. Sci.* 11:87-120; H. W. Schütt (1992). *Eilhard Mitscherlich: Baumeister am Fundament der Chemie.* Munich: Oldenbourg (English translation, 1997). See I. Freund (note Ch.3:9), pp. 411-427.

12. J. G. Burke (1966) (note Ch.2:51), pp. 122-131; D. C. Goodman (1969). Problems in crystallography in the nineteenth century. *Ambix* 16:152-166.

13. J. Dumas (1832). Dissertation sur la densité de la vapeur de quelques corps simples. *Ann. Chim.* [2] 50:170-181 (170).

14. W. Wollaston (1814). A synoptic scale of chemical equivalents. *Phil. Trans.* 104:1-23.

15. J. J. Berzelius (1819) (note Ch.5:7), p. 98.

16. C. A. Russell (1963). The electrochemical theory of Berzelius. *Ann. Sci.* 19:117-142.

17. At first, Berzelius used the opposite convention, with oxygen as the most electro-positive element.

18. F. Wöhler and J. Liebig (1832). Untersuchungen über des Radikal der Benzoesäure. *Ann. Pharm.* 3:249-287.

19. Berzelius's letter is appended to the reprint in H. Kopp (ed.) (1891). *Ostwald's Klassiker* no. 22. Leipzig: Engelmann, and included in the translation by O. T. Benfey (1963). *Classics in the Theory of Chemical Combination.* Pp. 15-39. New York: Dover.

20. J. J. Berzelius (1819) (note Ch.5:7), p. 45. Translation taken from C. A. Russell (1963) (note Ch.5:16), p. 137.

21. J. J. Berzelius (1819) (note Ch.5:7), pp. 95-96. Translation from C. A. Russell (1963) (note Ch.5:16), pp. 137-138.

22. Translated from B. S. Jørgensen (1964). Berzelius und die Lebenskraft. *Centaurus* 10:258-281 (275).

23. A. J. Rocke (1992) (note Ch.5:4); J. H. Brooke (1992). Berzelius, the dualistic hypothesis, and the rise of organic chemistry. In *Enlightenment Science etc.* (note Ch.5:4), pp. 180-221.

24. S. Lindroth (1992). Berzelius and his time. In *Enlightenment Science etc.* (note Ch.5:2), pp. 9-34.

25. F. Wöhler (1832). Über künstliche Bildung des Harnstoffs. *Pogg. Ann.* 12:253-256.

26. F. G. Hopkins (1928). The centenary of Wöhler's synthesis of urea. *Biochem. J.* 22:1341-1348; P. Walden (1928). Die Bedeutung der Wöhlerschen Harnstoff-Synthese. *Naturw.* 16:835-849; D. McKie (1944). Wöhler's synthetic urea and the rejection of vitalism: A chemical legend. *Nature* 153:608-610; J. Jacques (1950). Le vitalisme et la chimie organique pendant lapremière moitié du XIXe siècle. *Rev. Hist. Sci.* 3:32-66; T. O. Lipman (1964). Wöhler's preparation of urea and the fate of vitalism. *J. Chem. Educ.* 41:452-458; J. Schiller (1967). Wöhler, urée et le vitalisme. *Sudhoffs Arch.* 51:229-243; J. H. Brooke (1968). Wöhler's urea, and its vital force? A verdict from the chemists. *Ambix* 15:84-114.

27. A. W. Hofmann (1882). Friedrich Wöhler. *Ber. chem. Ges.* 15:1127-1190; J. Valentin (1949). Friedrich Wöhler. Stuttgart: Wissen-

schafliche Verlagsgesellschaft; R. Keen (1976). Friedrich Wöhler. *Dictionary of Scientific Biography* 14:474-479; G. Schwedt (1982). FriedrichWöhler zum 100. Todestag. *Naturw. Rund.* 35:406-413; H. Teichmann (1983). Zum Wirken Friedrich Wöhlers in Berlin. *Z. Chem.* 23:125-136.

28. Translated from J. Valentin (1949) (note Ch.5:27), p. 67.

29. A. W. Hofmann (ed. (1888). *Aus Justus Liebigs und Friedrich Wöhlers Briefwechsel in den Jahren 1829-1873*, vol. 1, p. 166. Braunschweig: Vieweg.

30. I. Freund (1904) (note Ch.3:9), pp. 545-572.

31. S. Kapoor (1969), Dumas and organic classification. *Ambix* 16:1-65; N. W. Fisher (1973). Organic classification before Kekulé. *Ambix* 20:106-131, 209-233.

32. J. J. Berzelius (1836). Einige Ideen über bei der Bildung organischer Verbindungen in der lebenden Natur wirksame, aber bisher nicht bemerkte Kraft. *Jahresber.* 15:237-245 (240, 243-245). See J. R. Partington (1964) (note Ch.5:9), pp. 261-264.

33. E. Buchner (1897). Alkoholische Gährung ohne Hefezellen. *Ber. chem. Ges.* 30:117-124.

34. J. J. Berzelius (1916). *Jac. Berzelius Bref* (H. G. Söderbaum, ed.). V. Correspondence between Berzelius and Mulder, 1834-1847, pp. 104-109 (106, 109). Uppsala: Almquist & Wiksell. See H. B. Vickery (1950). The origin of the word protein. *Yale J. Biol. Med.* 22:387-393 (389); H.A.M. Snelders (1952). The Mulder-Liebig controversy elucidated by their correspondence. *Janus* 69:199-221.

35. E. Fischer (1906). Untersuchungen über Aminosäuren, Polypeptide und Proteine. *Ber. chem. Ges.* 39:530-610.

36. A. E. Baudrimont (1833). Mémoire sur l'arrangement des atomes dans les molécules intégrantes. *J. Chim. Méd.* 9:39-41. See S. H. Mauskopf (1969). The atomic structural theories of Ampère and Gaudin: Molecular speculation and Avogadro's hypothesis. *Isis* 60:61-74; (1976). Crystals and compounds. *Trans. Am. Phil. Soc.* NS 66 (3), 1-82; J. A. Miller (1975). M. A. Gaudin and early nineteenth century stereochemistry. In *Van't Hoff-Le Bel Centennial* (O. B. Ramsey, ed.), pp. 1-17. Washington, D.C.: American Chemical Society; M. Scheidecker-Chevalier (1997). Alexandre-Edouard Baudrimont (1806-1880): Les liens entre sa chimie et sa philosophie. *Arch. Inter. Hist. Sci.* 47:27-56.

37. G. Lockemann (1949. *Robert Wilhelm Bunsen.* Stuttgart: Wis-

senschaftliche Verlags-gesellschaft. See H. Debus (1901). *Erinnerungen an Robert Wilhelm Bunsen.* Kassel: Fischer; T. Curtius and J. Bredt (1906). *Robert Bunsen als Lehrer in Heidelberg.* Heidelberg: Hörnung.

38. F.A.J.L. James (1985). The creation of a Victorian myth: The historiography of spectroscopy. *Hist. Sci.* 23:1-24; M. A. Sutton (1976). Spectroscopy and the chemists: a neglected opportunity? *Ambix* 23:16-26; I. D. Rae (1997). Spectrum analysis: The priority claims of Stokes and Kirchhoff. *Ambix* 44:131-143.

39. E. S. Eve and C. H. Creasey (1945). *Life and Work of John Tyndall.* p. 23. London: Macmillan.

40. R. Meyer (1904). Friedrich Knapp. *Ber. chem. Ges.* 37:4777-4814 (4781-4782).

41. J. R. Partington (1964) (note Ch.5:9), p. 300. Among the numerous biographical writings about Liebig are: J. Volhard (1909). *Justus von Liebig.* 2 vols. Leipzig: Barth; H. von Dechend (1965). *Justus von Liebig in eigenen Zeugnussen und solchen seiner Zeitgenossen.* Weinheim: Verlag Chemie; C. Paolini (1968). *Justus von Liebig, eine Bibliographie sämtlicher Veröffent-lichungen.* Heidelberg: Winter; F. L. Holmes (1973). Justus Liebig. *Dictionary of Scientific Biography* 8:329-350; P. Munday (1990). Social climbing through chemistry; Justus Liebig's rise from *niederer Mittelstand* to the *Bildungsbürgertum.* *Ambix* 37:3-19; W. H. Brock (1997). *Justus von Liebig: The Chemical Gatekeeper.* Cambridge University Press.

42. R. S. Turner (1982). Justus Liebig versus Prussian chemistry: reflections on early institute building in Germany. *Hist. Stud. Phys. Sci.* q13:129-162.

43. F. L. Holmes (1964). Introduction to *Animal Chemistry.* pp. vii-cxvi. New York: Johnson Reprint Corp.; M. W. Rossiter (1975). *The Emergence of Agricultural Science: Liebig and the Americans 1840-1880.* New Haven: Yale University Press; V.M.D. Hall (1980). The role of force or power in Liebig's physiological chemistry. *Med. Hist.* 24:20-59; M. R. Finlay (1991). The rehabilitation of an agricultural chemist: Justus von Liebig and the eighth edition. *Ambix* 38:155-167; P. Munday (1991). Liebig's metamorphosis from organic chemistry to the chemistry of agriculture. *Ambix* 38:135-154; U. Schling-Brodersen (1992). Liebig's role in the establishment of agricultural chemistry. *Ambix* 39:21-31; W. H. Brock (1997) (note Ch.5:41), pp. 145-214.

44. J. B. Morrell (1972). The chemist breeders: the research

schools of Liebig and Thomas Thomson. *Ambix* 19:1-46; W. Conrad (1985). Justus von Liebig und sein Einfluss auf die Entwicklung des Chemiestudiums und des Chemieunterrichts an Hochschulen und Schulen. PhD dissertation Darmstadt Technische Hochschule.

45. J. S. Fruton (1990). *Contrasts in Scientific Style*. Pp. 43-45. Philadelphia: American Philosophical Society.

46. P. Borscheid (1976). *Naturwissenschaft, Staat und Industrie in Baden 1848-1914*. Stuttgart: Klett; K. H. Jarausch (1982). *Students, Society, and Politics in Imperial Germany*. Princeton University Press.

47. Liebig's principal books, in the order of publication of the first edition: *Die organische Chemie in ihre Anwendung auf Agricultur und Physiologie* (1840); *Die Thierchemie oder die organische Chemie in ihre Anwendung auf Physiologie ind Pathologie* (1842); *Familiar Letters on Chemistry in Relation to Commerce, Physiology, and Agriculture* (1843); *Chemistry and Physics in Relation to Physiology and Pathology* (1846); *Die Grundsätze der Agricultur-Chemie, in Rücksicht auf die in angestellte Untersuchungen* (1855).

48. J. Liebig (1839). Über die Erscheinungen der Gärung, Fäulnis und Verwesung und ihre Ursachen. *Ann. Chem.* 30:250-288 (262).

49. J. Liebig (1840) (note Ch.5:47), p. 282.

50. E. Mitscherlich (1834). Ueber die Aetherbildung. *Ann. Phys.* 273-282 (273).

51. J. Liebig (1842). Mitscherlich und die Gärungschemie. *Ann. Chem.* 41:357-358.

52. T. S. Kuhn (1959). Energy conservation as an example of simultaneous discovery. In *Critical Problems in the History of Science* (M. Clagett, ed.), pp. 321-356. Madison, Wis.: University of Wisconsin Press.

53. J. Müller (1837). *Handbuch der Physiologie des Menschen*. Vol. 2, pp. 39-40. Coblenz: Hölscher. Translation from V.M.D. Hall (1980) (note Ch.5:43), p. 36. See also T. O. Lipman (1967). Vitalism and reductionism in Liebig's physiological thought. *Isis* 58:167-185; E. Benton (1974). Vitalism in nineteenth-century scientific thought; a typology and reassessment. *Stud. Hist. Phil. Sci.* 5:17-48; T. Lenoir (1982). *The Strategy of Life*. Dordrecht: Reidel.

54. M. Pelling (1978). *Cholera, Fever and English Medicine 1825-1865*. Pp, 133-145. Oxford University Press.

55. L. Pasteur (1871). Note sur un mémoire de M. Liebig relatif aux fermentations. *Compt. Rend.* 73:1419-1424.

56. K. E. Rothschuh (1975). Max Rubner. *Dictionary of Scientific Biography* 11:585-586; C. E. Rosenberg (1970). Wilbur Olin Atwater. Ibid., 1:325-326.

57. J. S. Fruton and S. Simmonds (1958). *General Biochemistry.* 2nd ed. pp. 928-938. K. J. Carpenter (1994). *Protein and Energy.* Cambridge University Press.

58. C. Bernard (1878-1879). *Leçons sur les Phénomènes de la Vie Communs aux Animaux et Végétaux.* Vol. 2, pp. 40-41. Paris: Baillière.

59. A. Fick and J. Wislicenus (1865). Ueber die Entstehung der Muskelkraft. *Vierteljahrschrift Züricher naturforschenden Gesellschaft* 10:317-348.

60. In 1846, Emil du Bois Reymond wrote his friend Karl Ludwig: "I consider his physiological theories as worthless and pernicious, because he entirely lacked the necessary factual knowledge and critical training." E. Du Bois Reymond (1927). *Zwei Grosse Naturforscher des 19. Jahrhunderts.* P. 19. Leipzig: Barth.

61. J. Liebig (1847). *Chemische Untersuchungen über das Fleisch und seine Zubereitung zum Nahrungsmittel.* Heidelberg: Winter; M. R. Finlay (1992). Quackery and cookery: Justus von Liebig's extract of meat and the theory of nutrition in the Victorian age. *Bull. Hist. Med.* 66:404-418.

62. G. R. Cowgill (1964). Jean Baptiste Boussingault. *J. Nutr.* 84:1-9; F.W.J. McCosh (1984). *Boussingault, Chemist and Agriculturist.* Dordrecht: Reidel; W. V. Farrar (1973). John Bennet Lawes and Joseph Henry Gilbert. *Dictionary of Scientific Biography* 8:92-93; E. J. Russell (1966). *A History of Agricultural Science in Great Britain.* London: Allen & Unwin.

63. J. B. Boussingault (1836, 1838). Recherches sur la quantité d'azote contenue dans les fourrages, et leur equivalents. *Ann. Chem.* [3] 63:225-244; 67:408-421; G. Ville (1854). Absorption de l'azote par les plantes. *Compt. Rend.* 38:705-709, 723-727; R. P. Aulie (1970). Boussingault and the nitrogen cycle. *Proc. Am. Phil. Soc.* 114:435-479; P. W. Wilson and R. H. Burris (1947). The mechanism of biological nitrogen fixation. *Bact. Revs.* 11:41-73.

64. R. P. Aulie (1974). The mineral theory. *Agricult. Hist.* 48:369-382.

65. The quotation is taken from V. M. Hall (1980) (note Ch.5:43), p. 59.

66. J. Liebig and J. B. Dumas (1837). Note sur l'état actuel de la chimie organique. *Compt. Rend.* 5:567-572.

67. J. R. Partington (1964) (note Ch.5:9), pp. 360-364.

68. J. B. Dumas (1840). Mémoire sur la loi des substitutions et la théorie des types. *Compt. Rend.* 10:149-178.

69. G. Söderbaum (1941). *Jac. Berzelius Brev* Supplement 2 (Correspondence betweeen Berzelius and Pelouze), pp. 58-67. Uppsala: Almquist & Wiksell.

70. R. Bunsen (1843). Mémoire sur l'acide cacodylique et sur le sulfure de cacodyle. *Ann. Chim.* [3] 8:356-362 (357).

71. D. C. Goodman (1972). Chemistry and the two kingdoms of nature in the nineteenth century. *Med. Hist.* 15:23-44.

72. J. B. Dumas and A. Cahours (1842). Mémoire sur les matières azotées neutres de l'organisation. *Ann. Sci. Nat.* 18:350-377.

73. H. Cassebaum (1971). Die Stellung der Arbeiten von J. B. Dumas (1800-1884) und A. Strecker (1822-1871) in der Entwicklung des Periodensystems. *NTM* 8:46-57. See also I. Freund(1904) (note Ch.3:9), pp. 454-505.

74. L. J. Klosterman (1985). A research school of chemistry in the nineteenth century: Jean Baptiste Dumas and his research students. *Ann. Sci.* 42:1-80.

75. C. de Milt (1951). Auguste Laurent, guide and inspiration of Gerhardt. *J. Chem. Educ.* 28:198-204; (1953). Auguste Laurent, founder of modern organic chemistry. *Chymia* 4:85-114; J. Jacques (1955). Essai bibliographique d'Auguste Laurent. *Arch. Inst. Grand-ducal de Luxembourg, Sect. Sci. Nat., Phys. Math.* NS 22:11-35; J. R. Partington (1964) (note Ch.5:9), pp. 364-393; S. C. Kapoor (1973). Auguste Laurent. *Dictionary of Scientific Biography* 8:54-61; M. Novitski (1992). *Auguste Laurent and the Prehistory of Valence.* Chur: Harwood; M. Blondel-Mégrelis (1996). *Dire les Choses: Auguste Laurent et la Méthode Chimique.* Paris: Vrin.

76. A. Laurent (1836). Théorie des combinaisons chimiques; quotation taken from C. de Milt (1953) (note Ch.5:75), p. 92.

77. A. Laurent (1837). Suites de recherches diverses de chimie organique. *Ann. Chim.* [3] 66:314-335; S. C. Kapoor (1969). The origins of Laurent's organic classification. *Isis* 60:477-527.

78. N. W. Fisher (1973) (note Ch.5:31), pp. 112-124. See also G. Bachelard (1973). *Le Pluralisme Cohérent de la Chimie Moderne,* pp. 58-63. Paris: Vrin.

79. J. B. Dumas (1838). Letter to Berzelius. *Compt. Rend.* 6:689-702 (699, 702). Quotation taken from C. de Milt (1953) (note Ch.5:75),

p. 98. See J. H. Brooke (1973). Chemical substitution and the future of organic chemistry: Methodological issues in the Laurent-Berzelius correspondence. *Stud. Hist. Phil. Sci.* 4:47-94.

80. J. B. Dumas (1840) (note Ch.5:68), p. 165.

81. J. B. Dumas (1857). Note sur les substitutions. *Ann. Chim.* [3] 57:487-496 (496).

82. J. Liebig (1838). Ueber Laurent's Theorie der organischen Verbindungen. *Ann. Pharm.* 25:1-31.

83. A. Laurent (1841. Recherches sur l'indigo. *Ann. Chim.* [4] 3:371-382.

84. L. E. Grimaux and C. Gerhardt (1900). *Charles Gerhardt, sa Vie, son Oeuvre, sa Correspondence.* Paris: Masson; M. Tiffeneau (1916). Conférence sur l'oeuvre de Charles Gerhardt. In Le Centenaire de Charles Gerhardt, pp. 13-105. *Bull. Soc. Chim.* Supplement; (1918). *Correspondence de Charles Gerhardt.* Paris: Masson; E. Kahane (1968). La vie et l'oeuvre de Charles Gerhardt. *Bull. Soc. Chim.* Pp. 4733-4742; M. P. Crosland and J. H. Brooke (1972). Charles Frédéric Gerhardt. *Dictionary of Scientific Biography* 5:369-375.

85. J. H. Brooke (1975). Laurent, Gerhardt, and the philosophy of chemistry. *Hist. Stud. Phys. Sci.* 6:405-429.

86. J. Liebig (1846), Herr Gerhardt und die organische Chemie. *Ann. Chem.* 57:389-394. See J. Volhard (1909) (note Ch.5:41), vol. 2, pp. 272-291.

87. Quotation from L. E. Grimaux and C. Gerhardt (1900) (note Ch.5:84), p. 201.

88. Ibid., p. 126.

89. C. Gerhardt (1853). Recherches sur les acides organiques anhydres. *Ann. Chim.* 57:285-342. See J. R. Partington (1964) (note Ch.5:9), p. 456-460.

90. C. Gerhardt (1838). Sur la constitution de l'alcool et ses dérivés. *J. prakt. Chem.* 15:17-54.

91. C. Gerhardt (1839). Recherches chimiques sur l'hellenine. *Ann. Chim.* [2] 72:163-183 (178).

92. N. W. Fisher (1973) (note Ch.5:31) pp. 215-217.

93. J. B. Dumas (1842). Loi de composition des principaux acids gras. *Compt. Rend.* 15:935-936.

94. C. Gerhardt (1843). Considérations sur les equivalents de quelques corps simples ou composés. *Ann. Chim.* [3] 8:238-245. See C. de Milt (1953) (note Ch.5:75), p. 106.

95. A. Laurent (1846). Sur les combinaisons azoteés. *Ann. Chim.* 18:266-298.

96. A. Laurent (1855). *Chemical Method.* pp. 7-12. London: Cavendish Society. See I. Freund (1904) (note Ch.3:9), pp. 194-201.

97. S. Cannizzaro (1947). *Sketch of a Course of Chemical Philosophy.* Edinburgh: Alembic Club. See W. A. Tilden (1912). Cannizzaro memorial lecture. *J. Chem. Soc.* 101:1677-1693; H. M. Leicester (1971). Stanislao Cannizzaro. *Dictionary of Scientific Biography* 3:45-49; J. Bradley (1992). *Before and after Cannizzaro.* Caithness: Wittles.

Chapter Six

1. J. R. Partington (1964) (note Ch.5:9), p. 437.

2. A. Laurent (1844). Classification chimique. *Compt. Rend.* 19:1089-1100 (1098-1099).

3. J. B. Dumas (1840) (note Ch.5:68), p. 156.

4. G. Söderbaum (1941) (note Ch.5:69).

5. S.C.H. Windler (1840). Über das Substitutionsgesetz und die Theorie der Typen. *Ann. Chem.* 33:108-110.

6. C. Gerhardt (1853) (note Ch.5:89).

7. A. Wurtz (1850). Mémoire sur une série d'alcaloides homologues avec ammoniaque. *Ann. Chim.* [3] 30:443-507; A. W. Hofmann (1850). Researches regarding the molecular constitution of the volatile organic bases. *Phil. Trans.*140:93-131; A. W. Williamson (1902). *Papers on Etherification and the Constitution of Salts.* Edinburgh: Alembic Club; E. Frankland (1850). Researches on organic radicals. *J. Chem. Soc.* 3:30-52.

8. A. Laurent (1855). *Chemical Method* (translated by W. Odling), p. 219. London: Cavendish Society.

9. Ibid., pp. 204-205; J. R. Partington (1964) (note Ch.5:9), pp. 372-375.

10. A. J. Rocke (1983). Subatomic speculations and the origin of the structural theory. *Ambix* 30:1-18; (1993). *The Quiet Revolution: Hermann Kolbe and the Science of Organic Chemistry.* Berkeley: University of California Press.

11. C. Friedel (1885). Notice sur la vie et des travaux de Charles Adolphe Wurtz. *Bull. Soc. Chim.* [2] 43:i-lxxx; A. Carneiro (1993). Adolphe Wurtz and the atomism controversy. *Ambix* 40:75-95; A. J.

Rocke (1993) (note Ch.6:10), pp. 96-98; (1994). Adolphe Wurtz and the development of organic chemistry in France: the Alsatian connection. *Bull. Soc. Indust. Mulhouse*, No. 2, pp. 29-34. A. Carneiro and N. Pigeard (1997). Chimistes alsaciens à Paris au 19ème siècle: un réseau, une école? *Ann. Sci.* 54:533-546.

12. An English translation by H. Watts was also published in 1869 under the title *A History of Chemical Theory from the Age of Lavoisier to the Present Time*. London: Macmillan.

13. A. J. Rocke (1993) (note Ch.6:10), pp, 101-107; M. Beugelmans-Verrier (1994). A propos des réactions de Wurtz. *Bull. Soc. Indust. Mulhouse*, No. 2, pp. 35-38; A. Wurtz and E. Bouchut (1879). Sur le ferment digestif du *Carica papaya. Compt. Rend.* 89:425-430; A. Wurtz (1880). Sur la papaine. Nouvelle contribution à l'étude des ferments solubles. Ibid., 91:787-791; (1885). *Traité de Chimie Biologique*. Paris: Masson.

14. N. Pigeard (1994). Un Alsatien à Paris: Charles Adolphe Wurtz (1817-1884), son école, ses laboratoires. *Bull. Soc. Indust. Mulhouse*, No. 2, pp. 39-43; A. J. Rocke (2001). *Nationalizing Science: Adolphe Wurtz and the Battle for French Chemistry*. Cambridge, Mass.: MIT Press.

15. E. C. Jungfleisch (1913). Notice sur la vie et les travaux de Marcellin *Berthelot. Bull. Soc. Chim.* [4] 13:i-cclx; L. Velluz (1964). *Vie de Berthelot*. Paris: Pion; J. Jacques (1987). *Berthelot 1827-1907: Autopsie d'un Mythe*. Paris: Belin. See H. W. Paul (1972). *The Sorcerer's Apprentice: The French Scientist's Image of German Science*. Gainesville: University of Florida Press; M. J. Nye (1981). Berthelot's anti-atomism: A matter of taste? *Ann. Sci.* 38:585-590.

16. G. K. Roberts (1976). The establishment of the Royal College of Chemistry: An investigation of the social context of early-Victorian chemistry. *Hist. Stud. Phys. Sci.* 7:437-485; J. Bentley (1970). The chemical department of the Royal School of Mines: Its origins and development under A. W. Hofmann. *Ambix* 17:153-181; (1972). Hofmann's return to Germany from the Royal College of Chemistry. *Ambix* 19:197-203.

17. L. Playfair, F. A. Abel, W. H. Perkin, and H. E. Armstrong (1896). Hofmann Memorial Lecture. *J. Chem. Soc.* 69:575-732; J. Volhard and E. Fischer (1902). *August Wilhelm Hofmann, ein Lebensbild*. Berlin: Friedländer; B. Lepsius (1930). A. W. von Hofmann. In *Das Buch der Grossen Chemiker* (G. Bugge, ed.), vol. 2, pp. 136-153; W. H. Brock (ed.) (1984). *Justus von Liebig und August Wilhelm Hofmann in*

ihren Briefen (1841-1873). Weinheim: Verlag Chemie; A. Etard (1892). Notice sur la vie et les travaux de Auguste Thomas Cahours. *Ann. Chim.* [6] 7:i-xii.

18. H. E. Armstrong. In L. Playfair et. al. (1896) (note Ch.6:17), pp. 640-641.

19. W. H. Brock (ed.) (1984) (note Ch.6:17), pp. 12-18.

20. B. Lepsius (1930) (note Ch.6:17), p. 144.

21. J. Harris and W. H. Brock (1974). From Giessen to Gower Street: Towards a biography of Alexander William Williamson (1821-1894). *Ann. Sci.* 31:95-130.

22. A. Williamson (1902). *Papers on Etherification and on the Constitution of Salts*, p. 8. Edinburgh: Alembic Club; J. R. Partington (1964) (note Ch.5:9), pp. 446-453.

23. A. Williamson (1902) (note Ch.6:22), pp. 43-44.

24. A. Williamson (1869). *J. Chem. Soc.* 22:328-365. Reprinted in D. M. Knight (ed.) (1968). *Classical Scientific Papers—Chemistry*, pp. 260-297; E. R. Paul (1978). Alexander W. Williamson on the atomic theory: A study of nineteenth century British atomism. *Ann. Sci.* 35:17-31.

25. W. V. Farrar (1964). Sir B. C. Brodie and his calculus of chemical operations. *Chymia* 9:169-179; D. M. Dallas (1967). The chemical calculus of Sir Benjamin Brodie. In W. D. Brock (ed.) (1967) (note Ch.1:8), *The Atomic Debates*, pp. 31-90; D. C. Goodman (1970). Benjamin Collins Brodie, Jr. *Dictionary of Scientific Biography* 2:484-486.

26. A. Cayley (1874). On the mathematical theory of isomers. *Phil. Mag.* 47:444-446; J. J. Sylvester (1878). On the application of the new atomic theory to graphical representations of covariants of binary quantities. *Am. J. Math.* 1:64-125. See K. H. Parschall (1997). Chemistry through invariant theory. In *Experiencing Nature* (H. Theerman and K. Hunger (eds.), pp. 81-111. Dordrecht: Kluwer.

27. J. Lederberg (1965). Topological mapping of organic molecules. *Proc. Natl. Acad. Sci.* 53:134-139.

28. J. L. Thornton and A. Wiles (1956). William Odling (1829-1921). *Ann. Sci.* 12:288-295; C. A. Russell (1971). *The History of Valency.* pp. 92-93, 120. Leicester University Press; W. H. Brock (1974). William Odling. *Dictionary of Scientific Biography* 10:177-179; S. Fisher (1996). William Odling: "interpreter and liaison officer." Advocate of a new system of chemistry. *Ambix* 45:146-163.

29. C. A. Russell (1986). *Lancastrian Scientist: The Early Years of Sir Edward Frankland.* Milton Keynes: Open University Press; (1996).

Edward Frankland: Chemistry, Controversy and Conspiracy in Victorian England. Cambridge University Press.

30. As in the case of Bunsen's "cacodyl," Frankland's "ethyl" turned out to be a dimeric product, in this case butane. This was clearly demonstrated by Carl Schorlemmer in 1864.

31. E. Frankland (1852). On a new series of organic bodies containing metals. *Phil. Trans.* 142:417-444.

32. A. Kekulé (1858). Ueber die Konstitution und Metamorphosen der chemischen Verbindungen und über die chemische Natur des Kohlenstoffs. *Ann. Chem.* 106:129-152. For an English translation, see O. T. Benfey (1963). *Classics in the Theory of Chemical Combination*, pp. 109-131. New York: Dover. See also N. W. Fisher (1974). Kekulé and organic classification. *Ambix* 21:29-52.

33. C. A. Russell (1971) (note Ch. 6:28), pp. 108-134; (1996) (note Ch.6:29), pp. 118-146; D. H. Rouvray (1992). Some key historical highlights in the evolution of the concept of valence. *J. Mol. Struct.* 259:1-28.

34. E. Frankland (1866). On the origin of muscular power. *Phil. Mag.* [4] 32:182-199.

35. A. S. Couper (1933). *On a New Chemical Theory and Researches on Salicylic Acid*, p. 19. Edinburgh: Alembic Club. In 1858, Couper published longer versions of *On a New Chemical Theory* in English (*Phil. Mag.* [4] 16:194-116) and in French (*Ann. Chim.* [3] 53:469-489. See C. A. Russell (1971) (note Ch.6:28), pp. 71-80, 124-125.

36. A. Kekulé (1858) (note Ch.6:32). He used the "barred" symbols of Berzelius to denote C = 12, O = 16, S = 32. Soon after the publication of Couper's paper, there appeared a priority claim: A. Kekulé (1858). Remarques de M. A. Kekulé à l'occasion d'une note de M. Couper sur une nouvelle théorie chimique. *Compt. Rend.* 47:378. Couper did not reply to this note.

37. R. Anschütz (1909), Life and chemical work of Archibald Scott Couper. *Proc. Roy. Soc. Edin.* 29:193-273.

38. In addition to Anschütz (1909), see L. Dobbin (1934). The Couper quest. *J. Chem. Educ.* 11:331-338; O. T. Benfey (1971). Archibald Scott Couper. *Dictionary of Scientific Biography* 3:448-450; D. G. Duff (1987). A. S. Couper: the forgotten genius. *Chem. Brit.* 23:350-354.

39. A. Kekulé (1890). Rede von August Kekulé. *Ber. chem. Ges.* 23:1302-1311. Reprinted in R. Anschütz (1929), *August Kekulé*, vol. 2, pp. 937-952. Berlin: Verlag Chemie.

40. See J. S. Fruton (1990) (note Ch.5:45), p. 22, note 17.

41. R. Anschütz (1929) (note Ch.6:39), vol. 2, p. 950.

42. A. Kekulé (1854). On a new series of sulphuretted acids. *Proc. Roy. Soc.* 7:37-40; (1864). Sur l'atomicité des éléments. *Compt. Rend.* 58:510-514.

43. R. Anschütz (1929) (note Ch.6:39), pp. 943-944.

44. A. Kekulé (1859). *Lehrbuch der organischen Chemie.* Erlangen: Enke. Subsequent installments were published in 1861, 1866, 1882, and 1887.

45. E. N. Hiebert (1959). The experimental basis of Kekulé's valence theory. *J. Chem. Educ.* 320-327.

46. J. Gillis (1961). Kekulé's life at Ghent (1858-1867). *J. Chem. Educ.* 38:118-122; (1966). Auguste Kekulé et son oeuvre, réalisée à Gand de 1858 à 1867. *Mem. Acad. Roy. Belg.* 37:1-40.

47. H. W. Schütt (1975). Guglielmo Körner (1839-1925) und sein Beitrag zur Chemie isomerer Benzolderivate. *Physis* 17:113-125.

48. A. J. Rocke (1985). Hypothesis and experiment in the early development of Kekulé's benzene theory. *Ann. Sci.* 42:355-381. See also A. F. Holleman (1915). Fünfzigjährisches Benzolstudium. *Janus* 20:459-488; A. Sementsov (1966). Who proposed the Dewar formula of benzene? *J. Chem. Ed.* 43:151.

49. W. Böhm (1973). Johann Joseph Loschmidt. *Dictionary of Scientific Biography* 8:507-511; C. R. Noe and A. Bader, Josef Loschmidt. In J. H. Wotiz (ed.) (1993). *The Kekulé Riddle: A Challenge for Chemists and Psychologists.* pp. 222-245. Clearwater: Cache River Press. A. J. Rocke (1993). Waking up to the facts? *Chem. Brit.* 29:401-402; W. Fleischhacker and T. Schönfeld (eds.) (1997). *Pioneering Ideas for the Physical and Chemical Sciences.* New York: Plenum.

50. A. Kekulé (1872). Ueber einige Condensations producte des Aldehyds. *Ann. Chem.* 162:72-128; A. Baeyer (1870). Ueber die Wasserentziehung und ihre Bedeutung für das Pflanzenleben und die Gährung. *Ber. chem. Ges.* 3:63-75.

51. G. F. Schiemenz (1993). A heretical look at the Benzolfest. *Brit. J. Hist. Sci.* 26:195-205.

52. R. Anschütz (1929) (note Ch.6:39), vol. 2, pp. 941-942. This approximately literal translation differs somewhat from that of other writers.

53. J. H. Wotiz and S. Rudofsky (1987). The unknown Kekulé. In *Essays on the History of Organic Chemistry* (J. G. Traynham, ed.), pp. 21-34. New York: American Book Co.; J. H. Wotiz (ed.) (1993). *The*

Kekulé Riddle: A Challenge for Chemists and Psychologists. Clearwater: Cache River Press.

54. A. J. Rocke (1985) (note Ch.6:48), p. 356.

55. S. F. Rudofsky and J. H. Wotiz (1988). Psychologists and the dream accounts of August Kekulé. *Ambix* 35:31-38 (38).

56. R. Anschütz (1929) (note Ch.6:39), vol 1, p. 133.

57. R. Anschütz (1929) (note Ch.6:39), vol. 2, pp. 183-209; C. A. Russell (1971) (note Ch.6:28), pp, 92-107; A. J. Rocke (1984). *Chemical Atomism in the Nineteenth Century*, pp. 287-311. Columbus: Ohio State University Press.

58. D. F. Larder (1967). Alexander Crum Brown and his doctoral thesis of 1861. *Ambix* 14:112-132; H. S. Mason (1943). History of the use of graphic formulas in organic chemistry. *Isis* 34:346-354.

59. See note Ch.6:49.

60. A. J. Rocke (1981). Kekulé, Butlerov, and the historiography of the theory of chemical structure. *Brit. J. Hist. Sci.* 14:27-57.

61. Quoted from H. M. Leicester (1959). Contributions of Butlerov to the development of structural theory. *J. Chem. Ed.* 36:328-329. See also H. M. Leicester (1940). Alexander Mikhailovich Butlerov. *J. Chem. Ed.* 17:203-209; G. V. Bykov (1962). The origin of the theory of chemical structure. *J. Chem. Educ.* 39:220-224.

62. Quoted from E. N. Hiebert (1959). The experimental basis of Kekulé's valence theory. *J. Chem. Educ.* 36:320-327 (324).

63. For the definitive account of Kolbe's life, work, and thought, see A. J. Rocke (1993) (note Ch.6:10).

64. A. J. Rocke (1993). Research groups and group research in German chemistry: Kolbe's Marburg and Leipzig institutes. *Osiris* [2] 8:51-79.

65. J. R. Partington (1962) (note Ch.5:9), pp. 516-532; N. W. Fisher (1973) (note Ch.5:31), pp. 222-231; A, J. Rocke (1993) (note Ch.6:10), pp. 181-209.

66. A. J. Rocke (1987). Kolbe versus the "transcendental chemists": the emergence of classical organic chemistry. *Ambix* 34:156-168.

67. H. Kolbe (1877). Zeichen der Zeit. *J. prakt. Chem.* 123:473-477; (1878). Kritik zur Rektoratrede von Aug. Kekulé etc. 125:139-163. The latter article includes an ironic riposte (pp. 159-163) from Kekulé.

68. H. Kolbe (1881). Meine Betheiligung an der Entwicklung der theoretischen Chemie. *J. prakt. Chem.* 131:305-323, 353-379, 497-517; 132:374-425.

69. Translation from above article (132:405) in A. J. Rocke (1993) (note Ch.6:10). p. 334.

70. Kekulé's paper was published in R. Anschütz (1929) (note Ch.6:39) vol. 1, pp. 540-569.

71. A. J. Rocke (1990). "Between two stools": Kopp, Kolbe, and the history of chemistry. *Bull. Hist. Chem.* 7:19-24; (1992-1993). Pride and prejudice in chemistry. Ibid., 13-14:29-40.

72. I. Freund (1904) (note Ch.3:9), pp. 454-505; J. W. Van Spronsen (1969). *The Periodic System of the Chemical Elements: a History of the First Hundred Years.* Amsterdam: Elsevier; D. C. Rawson (1974). The process of discovery: Mendeleev and the periodic law. *Ann. Sci.* 31:181-204; B. Bensaude-Vincent (1986). Mendeleev's periodic table of the elements. *Brit. J. Hist. Sci.* 19:3-17.

Chapter Seven

1. W. H. Wollaston (1923) (note Ch.4:19), p. 39.

2. M. E. Chevreul (1823). *Recherches Chimiques sur les Corps Gras de l'Origine Animale*, p. 2. Paris: Levrault.

3. Translation taken from Rocke (1981) (note Ch.6:60), p. 35. This statement precedes the one cited in note Ch.6:61.

4. A. Kekulé (1865). Sur la théorie atomique et la théorie de l'atomicité. *Compt. Rend.* 60:174-177 (174).

5. W. McGucken (1969). *Nineteenth-Century Spectroscopy: Development of the Understanding of Spectra.* Baltimore: Johns Hopkins University Press; M. A. Sutton (1976). Spectroscopy and the chemists: A neglected opportunity? *Ambix* 23:16-26; F.A.J.L. James (1983). The establishment of spectrochemical analysis as a practical method of qualitative analysis 1854-1861. *Ambix* 30:30-53.

6. T. M. Lowry (1935). *Optical Rotatory Power.* London: Longmans, Green; F. J. Bates (1942). *Polarimetry, Saccharimetry and the Sugars.* Washington: U.S. Govt. Printing Office; R. E. Lyle and G. G. Lyle (1964). A brief history of polarimetry. *J. Chem. Educ.* 41:308-313.

7. E. Picard (1931). La vie et l'oeuvre de Jean Baptiste Biot. In *Éloges et Discours Académiques*, pp. 221-287. Paris: Gauthier-Villars; M. P. Crosland (1970). *Dictionary of Scientific Biography* 2:133-140.

8. S. H. Mauskopf (1969). The atomic structural theories of Ampère and Gaudin: Molecular speculation on Avogadro's hypothesis. *Isis* 60:61-74.

9. J.F.W. Herschel (1822). On the rotation impressed on plates of rock crystal on the planes on polarization of the rays of light, in connection with certain peculiarities in its crystallization. *Proc. Camb. Phil. Soc.* 1:43-52.

10. Quotation taken from S. H. Mauskopf (1976). Crystals and compounds. *Trans. Am. Phil. Soc.* NS 66, part 3, pp. 1-82. (55).

11. Quotation taken from S. H. Mauskopf (1976) (note Ch.7:10), p. 71.

12. L. Pasteur (1848). Recherches sur le dimorphisme. *Compt. Rend.* 26:353-355 (355).

13. L. Pasteur (1922). *Oeuvres de Pasteur.* Vol 1, p. 78. Paris: Masson.

14. L. Pasteur (1861). Recherches sur la dissymétrie moléculaire des produits organiques naturels. In *Leçons de Chimie Professées en 1860.* pp. 1-48. Paris: Hachette. Translations into German and English have appeared in Ostwalds Klassiker 28 and Alembic Club reprint 14.

15. The best of the older biographies is that of E. Duclaux (1896). *Pasteur, Histoire d'un Esprit.* Sceux: Charaire. A more recent one is by R. Dubos (1950). *Louis Pasteur: Free Lance of Science.* Boston: Little Brown. For discussion of Pasteur's work on molecular dissymmetry, see G. B. Kauffman and R. D. Meyers (1975). The resolution of racemic acid, *J. Chem. Educ.* 52:777-781; S. H. Mauskopf (1976) (note Ch.7:10), pp. 65-80; D. B. Kottler (1978). Louis Pasteur and molecular dissymmetry. *Stud. Hist. Biol.* 2:57-98.

16. J. D. Bernal (1939). *Science and Industry in the Nineteenth Century,* pp. 181-219. London: Routledge & Kegan Paul; G. L. Geison and J. A. Secord (1988). Pasteur and the process of discovery: the case of optical isomerism. *Isis* 79:6-36; G. L. Geison (1995). *The Private Life of Louis Pasteur,* pp. 53-89. Princeton University Press.

17. J. H. van't Hoff and C. M. van Deventer (1887). Die Umwandlungstemperatur bei chemischer Zersetzung. *Z. physik. Chem.* 1:165-185.

18. Quotation taken from S. H. Mauskopf (1976) (note Ch.7:10), pp. 79-80.

19. See G. L. Geison (1995) (note Ch.7:16).

20. L. Pasteur (1922) (note Ch.7:13), vol. 1, pp. 327-328.

21. Ibid., vol. 1, p. 331.

22. L. Pasteur (1922) (note Ch.7:13), pp. 374-375.

23. Ibid., p. 380.

24. T. B. Osborne (1892). Crystallized vegetable proteins. *Am. Chem. J.* 14:662-689.

25. M. Traube (1878). Die chemische Theorie der Fermentwirkungen und der Chemismus der Respiration. *Ber. chem. Ges.* 11:1984-1992 (1984). See also M. Traube (1858). *Theorie der Fermentwirkungen.* Berlin: Dümmler.

26. F. A. Musculus (1876). Sur le ferment de l'urée. *Compt. Rend.* 82:333-336.

27. L. Pasteur (1933). Réponse à M. Berthelot. *Oeuvres de Pasteur.* Vol. 6, pp. 84-85 (85).

28. A French translation of the pamphlet: J. H. van't Hoff (1875). Sur les formules de structure dans l'espace. *Arch. Néerl. Sci. Ex. Nat.* 9:445-454; J. A. Le Bel (1874). Sur les relations qui existent entre les formules atomiques des corps organiques, et le pouvoir rotatoire de leurs dissolutions. *Bull. Soc. Chim.* [2] 22:337-347. English translations of both articles have been provided in G. M. Richardson (ed.) (1901). *The Foundations of Stereochemistry,* pp. 35-60. New York: American Book Co.; O. T. Benfey (ed.) (1963). *Classics in the Theory of Chemical Combination,* pp. 151-171. New York: Dover. Other books: J. H. van't Hoff (1875). *La Chimie dans l'Espace.* Rotterdam: Bazendijk; (1877). *Die Lagerung der Atome im Raume.* Braunschweig: Vieweg (2nd ed. 1894; 3rd ed. 1908); (1898). *The Arrangement of Atoms in Space.* London: Longmans, Green.

29. H.A.M. Snelders (1974). The birth of stereochemistry: An analysis of the 1874 papers of J. H. van't Hoff and J. A. Le Bel. *Janus* 60:261-278; E. Fischmann (1985). A reconstruction of the first experiments in stereochemistry. *Janus* 72:131-156; R. B. Grossman (1989). van't Hoff, Le Bel, and the development of stereochemistry: a reassessment. *J. Chem. Educ.* 66:30-33.

30. M. Delépine (1949). *Vie et Oeuvres de Joseph Achille Le Bel.* Paris: Dupont; P. Federlin (1974). Joseph Achille Le Bel (1847-1930), industriel, ingénieur et chomiste d'avant garde. *Saisons d'Alsace* NS 52: 84-102; (1994). J. A. Le Bel: Les pétroles du Bas-Rhin à la théorie de la dissymétrie moléculaire. *Bull. Soc. Ind. Mulhouse* No. 833, pp. 97-102.

31. J. P. Vigneron (1974). Le Bel, théoricien de l'assymétrie moléculaire. *Saisons d'Alsace* NS 52:103-122; W. J. Pope (1930). Joseph Achille Le Bel. *J. Chem. Soc.* pp. 2789-2792.

32. The definitive biography is E. Cohen (1912). *Jacobus Henricus van't Hoff: Sein Leben und Wirken.* Berlin: Akademische Verlagsgesellschaft; W. P. Jorissen and L. T. Reicher (1912). *J. H. van't Hoffs Amsterdamer Periode (1877-1895).* Den Helder: De Boer; H.A.M. Snelders (1984). J. H. van't Hoff's research school in Amsterdam

(1877-1895). *Janus* 71:1-30. See also P. J. Ramberg and G. J. Somsen (2001). Young J. H. van't Hoff: The background to the publication of his 1874 pamphlet etc. *Ann. Sci.* 58:51-74.

33. An English translation has been provided by O. T. Benfey (1960). The role of imagination in science. *J. Chem. Educ.* 37:467-470.

34. Cohen (1912) (note Ch.7:32), p. 85. J. Wislicenus (1873). Ueber die isomeren Milchsäuren. *Ann.* 166:3-64; Ueber die optisch-active Milchsäure der Fleischflüssigkeit, die Paramilchsäure. Ibid., 167:302-346; Ueber die Aethylenmilchsäure. Ibid., 167:346-356. See N. W. Fisher (1975). Wislicenus and lactic acid: the chemical background of van't Hoff's hypothesis. In O. B. Ramsay (ed.) (note Ch.5:36), pp. 33-54.

35. J. Wislicenus (1873) (note Ch.7:34), p. 343.

36. E. Beckmann (1904). Johannes Wislicenus. *Ber. chem. Ges.* 37:4861-4946; P. J. Ramberg (1994). Johannes Wislicenus, atomism, and the philosophy of chemistry; *Bull. Hist. Chem.* 15/16:45-53; P. R. Jones (1997). The young Johannes Wislicenus in America. Ibid., 20:28-32.

37. R. Anschütz (1929) (note 325), vol. 2, p. 912. For van't Hoff's relationship to Kekulé, see E. Fischmann (1985). A reconstruction of the first experiments in stereochemistry. *Janus* 72:131-156.

38. R. Anschütz (1929) (note Ch.6:39), vol. 2, p. 366.

39. H. A. M. Snelders (1974). The reception of J. H. van't Hoff's asymmetric carbon atom. *J. Chem. Ed.* 51:2-7; (1975). Practical and theoretical objections to J. H. van't Hoff's 1874 stereochemical ideas. In O. B. Ramsey (ed.) (1975) (note Ch.5:36), pp. 55-65; J. Weyer (1974). A hundred years of stereochemistry—The principal development phases in retrospect. *Angew. Chem. (Int. Ed.)* 13:591-598; (1977). Die Aufname der van't Hoffschen Hypothese vom asymmetrischen Kohlenstoffatom (1874) in Deutschland. In *Medizin, Naturwissenschaften und Technik und das zweite Kaiserreich* (G. Mann and R. Winau), pp. 311-320. Göttingen: Vanderhoek & Rupprecht.

40. J. Wislicenus (1887). Ueber die räumliche Anordnung der Atome in organischen Molekülen und ihre Bestimmung in geometrischen isomeren Verbindungen. *Abh. Sächs. Akad. Wiss. Math.-Phys. Cl.* 14:1-77. For an English translation, see G. M. Richardson (ed.) (1901) (note Ch.7:28), pp. 61-132.

41. V. Meyer (1890). Die Ergebnisse und Ziele der stereochemischen Forschung. *Ber. chem. Ges.* 23:567-619; E. Fischer (1894). Synthesen in der Zuckergruppe II. *Ber. chem. Ges.* 27:3189-3232; K.

Freudenberg (1966). Emil Fischer and his contribution to carbohydrate chemistry. *Adv. Carb. Chem.* 21:1-38.

42. A. Ihde (1959). The unraveling of geometric isomerism and tautomerism. *J. Educ. Chem.* 36:330-336.

43. H. Sachse (1890). Ueber the geometrische Isomerien der Hexamethylenderivate. *Ber. Chem. Ges.* 23:1365-1366; E. Mohr (1918). Die Baeyersche Spannungstheorie und die Struktur des Diamanten. *J. prakt. Chem.* 98:315-353; C. A. Russell (1975). The origin of conformational analysis. In O. B. Ramsey (note Ch.5:36), pp. 154-178; O. B. Ramsey (1987), The early history and development of conformational analysis. In J. G. Traynham (note Ch.6:53), pp. 54-96. K. B. Wiberg (1986). The concept of strain in organic chemistry. *Angew. Chem. (Int. Ed.).* 25:312-322; D.H.R. Barton (1956) Stereochemistry. In *Perspectives in Organic Chemistry* (A. Todd, ed.), pp. 68-95. New York: Interscience.

44. A. Hantzsch and A. Werner (1890). Ueber die räumliche Anordnung der Atome in Stickstoffhaltigen Molekülen. *Ber. chem. Ges.* 23:11-30; G. B. Kauffman (1966). Foundation of nitrogen stereochemistry: Alfred Werner's inaugural dissertation. *J. Chem. Educ.* 43:155-165; (1972). The stereochemistry of trivalent nitrogen compounds: Alfred Werner and the controversy over the structure of oximes. *Ambix* 19:129-144; (1973). "Quinquevalent" nitrogen and the structure of ammonium salts: Contributions of Alfred Werner and others. *Isis* 64:78-95.

45. P. J. Ramberg (1995). Arthur Michael's critique of stereochemistry, 1887-1899. *Hist. Stud. Phys. Biol. Sci.* 26:89-138.

46. J. A. Le Bel (1890). Sur les conditions d'équilibre des composées saturés de carbone. *Bull. Soc. Chim.* [3] 3:788. The English translation is taken from A. Severtsov (1955). The eightieth anniversary of the asymmetrical carbon atom. *Amer. Sci.* 43:97-100.

47. V. Prelog (1975). From configurational notation of stereoisomers to the conceptual basis of stereochemistry. In O. B. Ramsey (note Ch.5:36), pp. 179-187 (186); R. S. Cahn, C. Ingold, and V. Prelog (1966). Specification of molecular chirality. *Angew. Chem. (Int. Ed.)* 5:385-415. See I. Hargittai and M. Hargittai (1986). *Symmetry through the Eyes of a Chemist.* Weinheim: VCH; E. Heilbronner and J. D. Dunitz (1993). *Reflections on Symmetry.* Weinheim: VCH.

48. Quoted from S. F. Mason (1989). The development of concepts of chiral discrimination. *Chirality* 1:183-191 (189).

49. G. B. Kauffman (1966). *Alfred Werner, Founder of Coordination*

Chemistry. Berlin: Springer. See also A. Carneiro and N. Pigeard (1997). Chimistes alsaciens à Paris au 19e siècle: un réseau, une école? *Ann. Sci.* 54:533-546.

50. A. Werner (1890). Beiträge zur Theorie der Affinität und Valenz. *Viert. Zür. Nat. Ges.* 36:129-169. Quotation taken from English translation by G. B. Kauffman (1967). *Chymia* 12:189-216 (191-192).

51. G. B. Kauffman (1974). Early theories of metal-ammines. *J. Chem. Educ.* 51:522-524.

52. A. Werner (1893). Beitrag zur Konstitution anorganischer Verbindungen. *Z. anorg. Chem.* 31:267-330. An English translation has been provided by G. B. Kauffman (1968). *Classics in Coordination Chemistry*, vol. 1, pp. 5-88. New York: Dover.

53. A. Werner (1911). Zur Kenntnis des asymmetrischen Kobaltatoms I. *Ber. chem. Ges.* 44:1887-1898. An English translation was provided by G. B. Kauffman (1968) (note Ch.7:49), pp. 159-173. See also letter of V. L. King (1942). *J. Chem. Educ.* 19:241 and G. B. Kauffman (1975). The first resolution of a coordination compound. In O. B. Ramsey (ed.) (note Ch.5:36). pp. 126-142.

54. G. B. Kauffman (1960). Sophus Mads Jørgensen and the Werner-Jøregensen controversy. *Chymia* 6:180-204; H. Kragh (1997). S. M. Jørgensen and his controversy with A. Werner: a reconsideration. *Brit. J. Hist. Sci.* 30:203-219.

55. E. Fischer (1894). Einfluss der Konfiguration auf die Wirkung der Enzyme. *Ber. chem. Ges.* 27:2985-2993.

56. S. F. Mason (1987). From molecular morphology to universal dissymmetry. In J. G. Traynham (ed.) (note Ch.6:53), pp. 35-53; (1991). From Pasteur to parity violation: Cosmic dissymmetry and the origins of biomolecular handedness. *Ambix* 38:85-99; R. Bentley (1995). From optical activity in quartz to chiral drugs: Molecular handedness in biology and medicine. *Persp. Biol. Med.* 38:188-229.

57. J. H. Brooke (1971). Organic synthesis and the unification of chemistry—a reappraisal, *Brit. J. Hist. Sci.* 5:363-392; (1987). Methods and methodology in the development of organic chemistry. *Ambix* 34:147-155; C. A. Russell (1987). The changing role of synthesis in organic chemistry. *Ambix* 34:169-180.

58. A. Baeyer (1905). Erinnerungen aus meinem Leben. In *Adolf von Baeyer's Gesammelte Werke*, pp. vii-xx. Braunschweig: Vieweg; R. Willstätter et. al. (1915). Adolf von Baeyer zur Feier seines 80. Geburtstages. *Naturwissenschaften* 3:559-599; H. Rupe (1932). *Adolf von Baeyer als Lehrer*

und Forscher. Stuttgart: Enke; K. Schmorl (1952). *Adolf von Baeyer.* Stuttgart: Wissenschaftliche Verlagsgesellschaft; R. Huisgen (1986), Adolf von Baeyer's scientific achievements and legacy. *Angew. Chem. (Int. Ed.)* 25:297-311, J. S. Fruton (1990) (note Ch.5:45), pp. 118-162.

59. Quoted from W. H. Perkin (1933). Baeyer memorial lecture. In *Memorial Lectures Delivered before the Chemical Society 1914-1932.* pp. 47-73 (71). London: Chemical Society.

60. A. Baeyer (1870). Ueber die Wasserentziehung und ihre Bedeutung für das Pflanzenleben und die Gährung. *Ber. chem. Ges.* 3:63-75; (1872). *Ueber den Kreislauf des Kohlenstoffs in der organischen Natur.* Berlin: Lüderitz.

61. W. H. Perkin (1933) (note Ch.7:59), p. 68.

62. Ibid., pp. 72-73.

63. H. Rupe (1932) (note Ch.7:58), p. 24.

64. C. Harries (1919). Emil Fischers wissenschaftliche Arbeiten. *Naturwissenschaften* 7:843-860; K. Hoesch (1921). *Emil Fischer, sein Leben und sein Werk.* Berlin: Verlag Chemie; M. O. Forster (1923). Emil Fischer memorial lecture. (note Ch.7:59), pp. 1-45, H. Remane (1984). *Emil Fischer.* Leipzig: Teubner; J. S. Fruton (1990) (note Ch.5:45), pp. 163-229.

65. F. W. Lichtenthaler (1992). Emil Fischer's proof of the configuration of sugars: a centennial tribute. *Angew. Chem. (Int. Ed.)* 31:1541-1556.

66. Letters from Fischer to Baeyer, December 5, 1905 and April 18, 1906. (Bancroft Library).

67. E. Fischer (1907). Synthetical chemistry in relation to biology. *J. Chem. Soc.* 91:1749-1765 (1761, 1762).

68. G. D. Feldman (1973). A German scientist between illusion and reality. Emil Fischer 109-1919. In *Deutschland in der Weltpolitik des 19. und 20. Jahrhunderts.* (E. Geiss and B. J. Wendt, eds.), pp. 341-362. Düsseldorf: Bertelsmann; T. D. Moy (1989). Emil Fischer as "chemical mediator": Science, industry, and government during World War One. *Ambix* 36:109-120.

69. S. Olsen (1962). Otto Diels. *Chem. Ber.* 95:v-xlvi (xiv).

70. F. Herneck (1970). Emil Fischer als Mensch und Forscher. *Z. Chem.* 41-48 (45).

71. J. B. Herrick (1949). *Memories of Eighty Years.* pp. 131-132. University of Chicago Press.

72. M. Bergmann (1930), Emil Fischer. In *Das Buch der grossen Chemiker* (G. Bugge, ed.), vol. 2, pp. 408-420 (415). Berlin: Verlag Chemie.

73. F. Herneck (1970) (note Ch.7:70), p. 44.

74. H. T. Clarke (1958). Impressions of an organic chemist in biochemistry. *Ann. Rev. Biochem.* 27:1-14 (2).

75. T. Curtius and J. Bredt (1912). Wilhelm Koenigs. *Ber. chem. Ges.* 45:3781-3830.

76. W. Koenigs (1903). Hans von Pechmann. Ibid., 36:4417-4511.

77. R. Anschütz (1936). Ludwig Claisen. Ibid., 69A:97-170.

78. L. Blangey (1933). Eugen Bamberger. *Helv. Chim. Acta* 16:644-685.

79. A. Darapsky (1930). Theodor Curtius zum Gedächtnis. *J. prakt. Chem.* NF 125:1-22; K. Freudenberg (1963). Theodor Curtius. *Chem. Ber.* 96:i-xxv.

80. F. Straus (1927). Johannes Thiele. *Ber. chem. Ges.* 60A:75-132.

81. R. Robinson (1953). Willstätter memorial lecture. *J. Chem. Soc.*, pp. 999-1020.

82. L. Bert (1941). Otto Dimroth. *Ber. chem. Ges.* 74A:1-23.

83. E. Dane et al. (1942). Heinrich Wieland zu seinem 65. Geburtstag etc. *Naturwissenschaften* 30:333-359; B. Witkop (1992). Erinnerungen an Heinrich Wieland. *Liebigs Ann. Chem.* pp. i-xxxii; Remembering Heinrich Wieland (1877-1957): Portrait of an organic chemist and founder of modern biochemistry. *Med. Res. Revs.* 12:195-274.

84. H. Hopff (1959). Kurt H. Meyer. *Chem. Ber.* 92:cxxi-cxxxvi.

85. A. J. Greenaway et al. (1932). *The Life and Work of William Henry Perkin.* London: Chemical Society; J. Morrell (1993). W. H. Perkin, Jr., at Manchester and Oxford: From Irwell to Isis. *Osiris* 8:104-126.

86. J. C. Bailar (1970). Moses Gomberg. *Biog. Mem. Natl. Acad. Sci.* 41:141-173; J. M. McBride (1974). The hexaphenylethane riddle. *Tetrahedron* 30:2009-2022; R. Adams (1952). William Albert Noyes. *Biog. Mem. Natl. Acad. Sci.* 27:179-208. See P. R. Jones (1983). *Bibliographie der Dissertationen amerikanischer und britischer Chemiker an deutschen Universitäten, 1840-1914.* Munich: Deutsches Museum.

87. For a semi-popular account, see P. Ball (1993). *Designing the Molecular World.* Princeton University Press.

Chapter Eight

1. Z. Markovic (1970). Rudjer Bošković. *Dictionary of Scientific Biography* 2:326-332.

2. W. D. Brock and D. M. Knight (1965). The atomic debates:

"Memorable and interesting evenings in the life of the Chemical Society." *Isis* 56:5-25; H.A.M. Snelders (1971). Point-atomism in nineteenth-century Germany; M. J. Nye (1976). The nineteenth-century atomic debates and the dilemma of an 'indifferent hypothesis.' *Stud. Hist. Phil. Sci.* 7:245-268; D. K. Barkan (1992). A usable past: Creating disciplinary space for physical chemistry. In *The Invention of Physical Science*. pp. 175-202. Dordrecht: Kluwer.

3. E. N. Hiebert (1982). Developments in physical chemistry at the turn of the century. In *Science, Technology, and Society in the Time of Alfred Nobel* (C. G. Bernhard et al., eds.), pp. 97-115. Oxford: Pergamon; (1996). Discipline identifications in chemistry and physics. *Sci. Context* 9 (2):93-119.

4. W. R. Grove (1846). *The Correlation of Physical Forces*. London: Slipper and East; H. Helmholtz (1847). *Über die Erhaltung der Kraft*. Berlin: Reimer.

5. F. S. Taylor (1942). The origin of the thermometer. *Ann. Sci.* 5:129-156; C. B. Boyer (1943). History of the measurement of heat. *Sci. Mon.* 57:442-452, 546-554; D.S.L. Cardwell (1971). *From Watt to Clausius*. Manchester University Press.

6. M. Barnett (1958). Sadi Carnot and the second law of thermodynamics. *Osiris* 13:327-357; M. Kerker (1960). Sadi Carnot and the steam engine engineers. *Isis* 51:257-270; P. Costabel (1968). Le "calorique du vide" de Clément et Desormes. *Arch. Int. Hist. Sci.* 21:3-14; R. Fox (1970). Watt's expansive principle in the work of Sadi Carnot and Nicolas Clément. *Notes Roy. Soc.* 24:233-253; J. F. Challey (1971). Nicolas Léonard Sadi Carnot. *Dictionary of Scientific Biography* 3:79-84.

7. D.S.L. Cardwell (1983). The origins and consequences of certain of J. P. Joule's scientific ideas. In *Springs of Scientific Creativity* (R. Aris et al., eds.), pp. 44-70. Minneapolis: University of Minnesota Press; (1989). *James Joule*. Manchester University Press.

8. A. E. Musson and E. Robinson (1969). *Science and Technology in the Industrial Revolution*. Manchester University Press. See also F. Engels (1845). *Lage der arbeitenden Klasse in England*. Leipzig: Wigand. An English translation was prepared by W. O. Henderson and W. H. Chaloner (1968). The *Condition of the Working Class in England*. Stanford University Press.

9. R. H. Kargon (1977). *Science in Victorian Manchester*. Baltimore: Johns Hopkins University Press.

10. Quotation taken from Cardwell (1983) (note Ch.8:7), p. 53 (my

emphasis). See J. Forrester (1975). Chemistry and the conservation of energy: The work of James Prescott Joule. *Stud. Hist. Phil. Sci.* 6:274-313; W. H. Cropper (1988). James Joule's work in electrochemistry and the first law of thermodynamics. *Hist. Stud. Phys. Biol. Sci.* 19:1-15.

11. J. P. Joule (1845). On the existence of an equivalent relation between heat and the ordinary forms of mechanical power. *Phil. Mag.* [3] 27:205-207. See H. J. Steffens (1979). *James Prescott Joule and the Concept of Energy.* New York: Neale Watson.

12. J. D. Buchwald (1976). William Thomson. *Dictionary of Scientific Biography* 13:374-388; C. W. Smith and M. N. Wise (1989). *Energy and Empire: A Biographical Study of Lord Kelvin.* Cambridge University Press.

13. J. R. Mayer (1842). Bemerkungen über die Kräfte in der unbelebten Natur. *Ann. Chem.* 42:233-241; (1845). *Die organische Bewegung in ihrem Zusammenhange mit dem Stoffwechsel.* Heilbronn: Wechsler. See R. L. Kremer (1990). *The Thermodynamics of Life and Experimental Physiology*, pp. 215-231. New York: Garland; F. Kober (1982). Das Leben und Werk des Julius Robert Mayer. *Chemiker-Zeitung* 106:397-405; K. L. Caneva (1993). *Robert Mayer and the Conservation of Energy.* Princeton University Press. Mayer's published papers were collected by J. J. Weyrauch (1893), *Die Mechanik der Wärme.* 3rd ed. Stuttgart: Cotta.

14. Letter from Joule to Thomson, 11 March 1851. Quoted from D.S.L. Cardwell (1989) (note 452), p. 126.

15. See D.S.L. Cardwell (1989) (note Ch.8:7), pp. 93-94, 280-281; F. Bevilacqua (1993). Helmholtz's *Ueber die Erhaltung der Kraft*: The emergence of a theoretical physicist. In *Hermann von Helmholtz and the Foundations of Nineteenth-Century Science* (D. Cahan, ed.), pp. 291-333 (321-323). Berkeley: University of California Press.

16. H. Helmholtz (1882). *Wissenschaftliche Abhandlungen.* Vol. 2, p. 79. Leipzig: Barth. This addendum is also included in the Ostwald's Klassiker series (no. 1), pp. 58-59.

17. L. Koenigsberger (1902-1903). *Hermann von Helmholtz.* 3 vols. Braunschweig: Vieweg; R. S. Turner (1972). *Dictionary of Scientific Biography* 6:241-253; D. Cahan (ed.) (1993) (note Ch.8:15); L. Krüger (ed.) (1994). *Universalgenie Helmholtz.* Berlin: Akademie Verlag.

18. H. Helmholtz (1845). Ueber den Stoffverbrauch bei der Muskelaction. *Arch. Anat. Physiol.*, pp. 72-83 (72). See K. M. Olesko and F. L. Holmes (1993). Experiment, quantification, and discovery: Helm-

holtz's early physiological researches. In D. Cahan (ed.) (note Ch.8:15), pp. 50-108 (54-59).

19. H. Helmholtz (1845) (note Ch.8:18), pp. 78, 82.

20. J. S. Fruton (1979). Early theories of protein structure. *Ann. N.Y. Acad. Sci.* 325:1-15.

21. Quoted from E. Cohen (1912) (note Ch.7:32), p. 314.

22. H. Helmholtz (1847) (note Ch.8:4), p. 17.

23. Ibid., p. 70. See R. L. Kremer (1990) (note Ch.8:13), pp. 248-255.

24. G. H. Hess (1840). Thermochemische Untersuchungen. *Ann. Phys.* 50:385-404 (392).

25. M. Rubner (1894). Die Quelle der thierischen Wärme. *Z. Biol.* 30:73-142; F. G. Benedict (1938). *Vital Energetics*. Washington, D.C.: Carnegie Institution.

26. J. Thomsen (1861). Om de chemiske Processers almindelige Characteer og en paa denne bygget Affinitetslære. *Arch. Chem.* 18:433-440, 481-495. Quoted from S. Veibel (1939). *Kemien i Danmark*. Vol. 1, p. 205. Copenhagen: Nordisk Verlag. See V. M. Schelar (1966). Thermochemistry and the third law of thermodynamics. *Chymia* 11:99-124; R.G.A. Dolby (1984). Thermochemistry versus thermodynamics. *Hist. Sci.* 32:375-400; H. Kragh (1984). Julius Thomsen and classical thermochemistry. *Brit. J. Hist. Sci.* 17:255-272; L. Médard and H. Tachoire (1994). *Histoire de la Thermochimie*. Aix-en-Provence: Université de Provence.

27. M. Berthelot (1873). Sur la statique des dissolutions salines. *Bull. Soc. Chim.* [2] 19:160-164 (160).

28. R. Clausius (1850). Ueber die bewegende Kraft der Wärme und die Gesetze welche sich daraus für die Wärmelehre selbst ableiten lassen. *Ann. Phys.* 79:368-397, 500-524. Reprinted in Ostwald's Klassiker No. 99.

29. F. Folie (1890). R. Clausius, sa vie, ses oeuvres et leurs portée metaphysique. *Rev. Quest. Sci.* 27:419-487; E. Daub (1971). Rudolf Clausius. *Dictionary of Scientific Biography* 3:303-311; S. L. Wolff (1995). Clausius' Weg zur kinetischen Gastheorie. *Sudhoffs Arch.* 79:54-72.

30. L. Clausius (1850) (note Ch.8:28), p. 383.

31. L. Clausius (1865). Ueber verschiedene für die bequeme Formen der Haupt-gleichungen der mechanischen Wärmetheorie. *Ann. Phys.* 125:353-400; (1868) On the second fundamental theorem of the mechanical theory of heat. *Phil. Mag.* [4] 35:405-419 (419).

32. E. E. Daub (1967). Atomism and thermodynamics. *Isis*

58:293-303; (1970). Entropy and dissipation. *Hist. Stud. Phys. Sci.* 2:321-354; K. Hutchison (1973). Der Ursprung der Entropiefunktion bei Rankine und Clausius. *Ann. Sci.* 30:341-364; (1981). W.J.M. Rankine and the rise of thermodynamics. *Brit. J. Hist. Sci.* 14:1-26.

33. S. G. Brush (1957). The development of the kinetic theory of gases. I. Herapath. *Ann. Sci.* 13:188-198; II. Waterston. Ibid., pp. 273-282; E. Mendoza (1975). A critical examination of Herapath's dynamic theory of gases. *Brit. J. Hist. Sci.* 8:155-165.

34. E. W. Garber (1972). Clausius' and Maxwell's kinetic theory of gases. *Hist. Stud. Phys. Sci.* 2:299-319; S. G. Brush (1976). The kinetic theory of gases. III. Clausius. *Ann. Sci.* 14:183-196.

35. M. J. Klein (1970). Maxwell, his demon, and the second law of thermodynamics. *Am. Sci.* 58:84-97; C.W.F. Everitt (1974). James Clerk Maxwell. *Dictionary of Scientific Biography* 9:198-230; (1983). Maxwell's scientific creativity. In Aris et al. (eds.) (note Ch.8:7), pp. 71-141. M. Goldman (1982). *The Demon in the Aether: The Story of James Clerk Maxwell.* Edinburgh: Harris.

36. S. G. Brush (1970). Ludwig Boltzmann. *Dictionary of Scientific Biography* 2:260-268; (1976). *The Kind of Motion we Call Heat.* Amsterdam: North-Holland. See also M. J. Klein (1969). Gibbs on Clausius. *Hist. Stud. Phys. Sci.* 1:127-149.

37. M. Trautz (1930). August Friedrich Horstmann. *Ber. chem. Ges.* 63A:61-86; A. Kipnis (1997). *August Friedrich Horstmann und die physikalische Chemie.* Berlin: ERS-Verlag.

38. A. F. Horstmann (1869). Dampfspannung und Verdampfungswärme des Salmiaks. *Ber. chem. Ges.* 2:137-140; (1873). Zur Theorie des Dissociation. *Ann. Chem.* 170:192-210 (193). See J. R. Partington (1964) (note Ch.5:9), p. 618; H. M. Snelders (1972). Dissociation, Darwinism and entropy. *Janus* 64:51-75; J. Berger (1997). Chemische Mechanik und Kinetik: die Bedeutung der mechanischen Wärmetheorie für die Theorie chemischer Reaktionen. *Ann. Sci.* 54:567-584.

39. W. Ostwald (1902). *Lehrbuch der allgemeinen Chemie.* 2nd ed., vol. 2, p. 114. Leipzig: Engelmann.

40. Rayleigh (1875). On the dissipation of energy. *Nature* 11:454-455. Quotation taken from H. Kragh and S. J. Weininger (1996). Sooner silence than confusion: The tortuous entry of entropy into chemistry. *Hist. Stud. Phys. Biol. Sci.* 27:91-130 (98).

41. W. L. Miller (1925). The method of Willard Gibbs in chemi-

cal thermodynamics. *Chem. Revs.* 1:293-344; M. J. Klein (1983). The
scientific style of Josiah Willard Gibbs. In R. Aris et al. (eds.) (note
Ch.8:7), pp. 142-162.

42. J. W. Gibbs (1906). *The Scientific Papers of J. Willard Gibbs* (H. A.
Bumstead and R. G. van Name, eds.) 2 vols. London: Longmans, Green.

43. P. Duhem (1908). Josiah-Willard Gibbs à propos de la publica-
tion de ses mémoires scientifiques. *Rev. Quest. Sci.* 63:5-43. See M. J.
Klein (1990). Duhem on Gibbs. In *Beyond the History of Science* (E.
Garber, ed.), pp. 52-66. Lehigh University Press.

44. L. P. Wheeler (1952). *Josiah Willard Gibbs. The History of a Great
Mind.* 2nd ed. New Haven: Yale University Press; M. J. Klein (1972).
Josiah Willard Gibbs. *Dictionary of Scientific Biography* 5:386-393.

45. J. W. Gibbs (1873). Graphical methods in the thermodynamics
of fluids. *Trans. Conn. Acad. Arts Sci.* 2:309-342; A method of geometrical representation of the thermodynamic properties of substances by
means of surfaces. Ibid., 2:382-404. See M. J. Klein (1989). The physics of J. Willard Gibbs in his time. In *Proceedings of the Gibbs Symposium* (D. G. Caldi and G. D. Mostow, eds.), pp. 1-21. Providence:
American Mathematical Society.

46. E. E. Daub (1976). Gibbs phase rule. A centenary retrospect.
J. Chem. Educ. 53:747-751. See also A. Y. Kipnis et al. (1996). *Van der
Waals and Molecular Science.* Oxford: Clarendon Press.

47. J. W. Gibbs (1902). *Elementary Principles of Statistical Mechanics
Developed with Special Reference to the Rational Foundations of Thermodynamics.* New York: Scribner's.

48. L. P. Wheeler (ed.) (1947). *The Early Work of Willard Gibbs in
Applied Mechanics.* New York: Shuman.

49. H. Helmholtz (1882). Die Thermodynamik chemischer
Vorgänge. *Sitz. Preuss. Akad. Wiss.*, pp. 23-29.

50. Ibid., (1883), pp. 647-655.

51. H.A.M. Snelders (1984) (note Ch.7:32).

52. J. H. van't Hoff (1877, 1881). *Ansichten über die organische Chemie.*
Braunschweig: Vieweg.

53. J. H. van't Hoff (1884). *Études de Dynamique Chimique.* Amsterdam: Muller.

54. J. H. van't Hoff (1898-1900). *Vorlesungen über theoretische und
physikalische Chemie.* Braunschweig: Vieweg.

55. F. L. Holmes (1962). From elective affinities to chemical equilibria. *Chymia* 8:105-145; M. W. Lindauer (1962). The evolution of

the concept of chemical equilibrium from 1775 to 1923. *J. Chem. Educ.* 39:384-390.

56. L. F. Wilhelmy (1850). Über das Gesetz nach welchem die Einwirkung der Säuren auf den Rohrzucker stattfindet. *Ann. Phys.* 81:413-433, 499-526; A. V. Harcourt and W. Esson (1865). On the laws of connexion between the conditions of a chemical change and its amount. *Phil. Trans.* 156:193-222; C. M. Guldberg and P. Waage (1867). *Études sur les Affinitées Chimiques.* Christiania: Brøgger and Christie (see Ostwald Klassiker No. 104). See K. J. Laidler (1985). Chemical kinetics and the origin of physical chemistry. *Arch. Hist. Exact Sci.* 32:43-75.

57. W. Pfeffer (1877). *Osmotische Untersuchungen.* Leipzig: Engelmann; J. H. van't Hoff (1887). Die Rolle des osmotischen Druckes in der Analogie zwischen Lösungen und Gasen *Z. physik. Chem.* 1:481-508.

58. E. Riesenfeld (1931). *Svante Arrhenius.* Leipzig: Akademische Verlagsgesellschaft; E. Crawford (1996). *Arrhenius.* Canton, MA: Science History Publications.

59. W. Ostwald (1884). Die elektrische Leitungsfähigkeit der Säuren. *J. prakt. Chem.* [2] 30:225-237.

60. R. Clausius (1857). Über die Elektricitätsleitung in Elektrolyten. *Ann. Phys.* 101:338-360; J. W. Hittorf (1859). Über die Wanderung der Ionen während der Elektrolyse. *Ann. Phys.* 106:337-411, 513-586; F. Kohlrausch and O. Grotrian (1875). Das elektrische Leitungs-vermögen der Chlor-Alkalien und alkalinischen Erden etc. *Ann. Phys.* 154:1-14; 215-239.

61. S. Arrhenius (1887). Über die Dissociation der in Wasser gelösten Stoffe. *Z. physik. Chem.* 1:631-648. An English translation appeared in the Alembic Club reprint No. 19 (1929).

62. S. Arrhenius (1889). Über die Reaktionsgeschwindigkeit bei der Inversion von Rohrzucker durch Säuren. *Z. physik. Chem* 4:226-248. See S. R. Logan (1982). The origin and status of the Arrhenius equation. *J. Chem Educ.* 59:279-281; K. J. Laidler (1984). The development of the Arrhenius equation. Ibid., 61:494-498.

63. E. Crawford (1984). *The Beginnings of the Nobel Institution. The Science Prizes, 1901-1915.* Cambridge University Press; (1996). *Arrhenius.* Canton, Mass.: Science History Publications.

64. L. P. Rubin (1980). Styles in scientific explanation: Paul Ehrlich and Svante Arrhenius on immunochemistry. *J. Hist. Med.* 35:397-425; J. S. Fruton (1982). The interplay of biology and chemistry at the turn of

the century. In C. G. Bernhard et al. (note Ch.8:3), pp. 91-94; F. Lüttenberger (1992). Arrhenius vs. Ehrlich on immunochemistry: Decisions about scientific progress in the context of the Nobel Prize. *Theor. Med.* 13:137-173; E. Crawford (1996) (note Ch.8:63), pp. 175-239.

65. P. Walden (1932). Wilhelm Ostwald. *Ber. Chem. Ges.* A65:101-141; F. G. Donnan (1933). Ostwald Memorial Lecture. *J. Chem. Soc.*, pp. 316-332; E. N. Hiebert and H. G. Körber (1978). Friedrich Wilhelm Ostwald. *Dictionary of Scientific Biography* 15:455-469; J. W. Servos (1990). *Physical Chemistry from Ostwald to Pauling.* Princeton University Press.

66. H. Kellermann (1915). *Das Krieg der Geister,* pp. 64-68. Weimar: Duncker; K. Schwabe (1969). *Wissenschaft und Kriegsmoral,* pp. 22-23. Göttingen: Musterschmidt.

67. W. Ostwald (1884). Studien zur chemischen Dynamik. *J. prakt. Chem.* 29:385-400.

68. W. Ostwald (1885). Elektrochemische Studien. II. Das Verdünnungsgesetz. *J. prakt. Chem.* 31:433-462. See J. H. Wolfenden (1972). The anomaly of strong electrolytes. *Ambix* 19:175-196.

69. R. Peters (1898). Über Oxydations- und Redukionsketten und den Einfluss komplexer Ionen auf ihre elektromotorishe Kraft. *Z. physik. Chem.* 26:193-236.

70. W. Ostwald (1900). Über Oxydationen mittels freien Sauerstoff. *Z. physik. Chem.* 34:248-252 (250).

71. W. Ostwald (1901). Über Katalyse. *Z. Elektrochem.* 7:995-1004. See K. J. Laidler (1986). Development of theories of catalysis. *Arch. Hist. Exact Sci.* 35:345-374.

72. See J. W. Servos (1990) (note Ch.8:65).

73. R.G.A. Dolby (1976). Debates over the theory of solutions: A study of dissent in physical chemistry in the English-speaking world in the late nineteenth and early twentieth centuries. *Hist. Stud. Phys. Sci.* 7:297-404.

74. W. Ostwald (1904). Faraday Memorial Lecture. *J. Chem. Soc.* 85:506-522.

75. Quotation taken from E. N. Hiebert and H. G. Körber (1978) (note Ch.8:65), p. 464.

76. A. Leegwater (1986). The development of Wilhelm Ostwald's chemical energetics. *Centaurus* 29:314-337; C. Hakfoort (1992). Science deified: Wilhelm Ostwald's energeticist world-view and the history of scientism. *Ann. Sci.* 49:525-544.

77. W. Ostwald (1926-1927). *Lebenslinien.* 3 vols., vol. 3, pp. 111-112. Berlin: Klasing.

78. M. Bodenstein (1942). Walther Nernst. *Ber. chem. Ges.* A75:79-104; W. Haberditzl (1960). Walther Nernst und die Tradition der physikalichen Chemie an der Berliner Universität. In *Forschen und Wirken*, pp. 401-416. Berlin: VEB Deutscher Verlag der Wissenschaften; K. Mendelsohn (1973). *The World of Walther Nernst.* University of Pittsburgh Press; E. N. Hiebert (1978). Hermann Walther Nernst. *Dictionary of Scientific Biography* 15:432-453; D. N. Barkan (1999). *Walther Nernst and the Transition to Modern Physical Science.* New York: Cambridge University Press.

79. W. Nernst (1889). Die elektromotorische Wirkung der Ionen. *Z. physik. Chem.* 4:129-181.

80. W. Nernst (1891). Über die Verteilung eines Stoffe zwischen zwei Lösungen und Dampfraum. Ibid., 8:110-139.

81. W. U. Eckart and K. Volkert (1996). *Hermann von Helmholtz.* Pp. 319-320. Pfafferweiler: Centaurus. Another part of this letter was cited above (see note Ch.8:21).

82. W. Nernst (1906). Über die Berechnung chemischer Gleichgewichte aus thermischen Messungen. *Nachr. Ges. Wiss. Göttingen*, pp. 1-40.

83. H. Le Châtelier (1888). Recherches expérimentales et théoriques sur les equilibres chimiques. *Annales des Mines et des Carburents.* 13:157-382; G. N. Lewis (1899). The development and application of a general equation for free energy and physico-chemical equilibrium. *Proc. Am. Acad. Arts Sci.* 35:3-38; T. W. Richards (1902). The significance of the change in atomic volumes. Ibid., 38:293-317; F. Haber (1905). *Thermodynamik technischer Gasreaktionen.* Munich: Oldenbourg.

84. M. J. Klein (1965). Einstein, specific heats, and the early quantum theory. *Science* 148:173-180.

85. J. Mehta (1975). *The Solvay Conference in Physics: Aspects of the Development of Physics since 1911.* Dordrecht: Reidel.

86. J. Báron and M. Polanyi (1913). Über die Anwendigkeit des zweiten Hauptsatzes der Thermodynamik auf Vorgänge im tierischen Organismus. *Biochem. Z.* 53:1-20.

87. W. Nernst (1904). Discussion after a lecture by Arrhenius. *Z. Elektrochem.* 10:676-677.

88. W. Nernst (1893). *Theoretische Chemie vom Standpunkt der Avo-*

gadroischen Regel und der Thermodynamik. Stuttgart: Enke; W. Nernst and A. Schönflies (1895). *Einführung in die mathematische Behandlung der Naturwissenschaften.* Munich: Lwoff.

Chapter Nine

1. The most valuable of the recent biographies are those of L. P. Williams (1965). *Michael Faraday.* New York: Basic Books, and J. M. Thomas (1991). *Michael Faraday and the Royal Institution.* Bristol: Adam Hilger. Older ones include those by J. Tyndall (1868), *Faraday as a Discoverer.* London: Longmans, Green; H. Bence Jones (1869). *The Life and Letters of Michael Faraday.* 2 vols. London: Longmans, Green; S. P. Thompson (1898). *Michael Faraday: His Life and Work.* New York: Macmillan. I will also refer to some of the articles in a special issue (no. 11, 1991) of the *Bulletin for the History of Chemistry.*

2. L. P. Williams (1971). Michael Faraday. *Dictionary of Scientific Biography* 4:527-540; J. M. Thomas (1991). Faraday the man, Faraday the genius. *Analyst* 116:1205-1210.

3. A more extreme interpretation has been offered by G. Cantor (1991). *Michael Faraday: Sandemanian and Scientist: A Study of Science and Religion in the Nineteenth Century.* New York: St. Martin's Press. See H. T. Pratt (1991). Michael Faraday's Bibles as mirrors of his faith. *Bull. Hist. Chem.* 11:40-47.

4. B. Bowers and L. Symons (1991). *Curiosity Perfectly Satisfied. Faraday's Travels in Europe 1813-1815.* London: Peter Peregrine.

5. J. Z. Fullmer and M. C. Usselman (1991). Faraday's election to the Royal Society: A reputation in jeopardy. *Bull. Hist. Chem.* 11:17-28.

6. D. A. Davenport (1991). Observations on Faraday as an organic chemist manqué. *Bull. Hist. Chem.* 11:60-65; L. C. Newell (1926). Faraday's discovery of benzene. *J. Chem. Educ.* 3:1248-1253; (1931). Faraday's contributions to chemistry. Ibid., 6:1493-1522.

7. W. J. Jensen (1991). Michael Faraday and the art and science of chemical manipulation. *Bull. Hist. Chem.*, pp. 65-76; S. Ross (1991). The chemical manipulator. Ibid., pp. 76-79.

8. F. Steinle (1995). Looking for a "simple case": Faraday and electromagnetic rotation. *Hist. Sci.* 33:179-202.

9. T. Martin (1949). *Faraday's Discovery of Electro-Magnetic Induction.* London: Edward Arnold; S. Ross (1965). The search for elec-

tromagnetic induction 1820-1831. *Notes Roy. Soc.* 20:184-219; J. Romo and M. G. Doncel (1994). Faraday's initial mistake concerning the direction of induced currents, etc. *Arch. Hist. Exact Sci.* 47:291-385; F. Steinle (1996). Work, finish, publish? The formation of the second series of Faraday's Experimental Researches in Electricity. *Physis* 33:141-220.

10. M. Faraday (1834). Experimental researches in electricity. 7th series. On electro-chemical decomposition. *Phil. Mag.* 5:77-122 (§783, 836). See J. T. Stock (1991). The pathway to the laws of electrolysis. *Bull. Hist. Chem.* 11:86-92.

11. S. Ross (1961). Faraday consults the scholars: The origins of the terms of electro-chemistry. *Notes Roy. Soc.* 16:187-220.

12. M. Faraday (1834). Experimental researches in electricity. 6th series. On the power of metals and other solids to induce the combustion of gases. *Phil. Trans.* 124:55-76. See P. Collins (1976) (note Ch.4:28).

13. M. Faraday (1838). Experimental researches in electricity. 11th series. On induction. *Phil. Mag.* 13:281-299, 355-367, 413-430.

14. M. Faraday (1844). A speculation touching electric conduction and the nature of matter. *Phil. Mag.* 24:136-144.

15. M. Faraday (1844). Liquefaction and solidification of bodies generally existing as gases. *Phil. Trans.* 135:155-177. (Alembic Club reprint No. 12).

16. M. Faraday (1846). Experimental researches in electricity. 19th series. On the magnetization of light and the illumination of magnetic lines of force. *Phil. Mag.* 28:294-317.

17. M. Faraday (1846). Experimental researches in electricity. 20th series. On new magnetic actions, and the magnetic conditions of all matter. *Phil. Mag.* 28:396-468.

18. It has been suggested that Faraday's medical problems were a consequence of mercury poisoning. J. F. O'Brien (1991). Faraday's health problems. *Bull. Hist. Chem.* 11:47-50.

19. G. Cantor (1991). Educating the judgment: Faraday as a lecturer. *Bull. Hist. Chem.* 11:28-36.

20. L. P. Williams (1991). Faraday and his biographers. *Bull. Hist. Chem.* 11:9-17.

21. J. C. Maxwell (1856). Faraday's lines of force. *Proc. Cambr. Phil. Soc.* 10:21-76.

22. C.W.F. Everitt (1974). James Clerk Maxwell. *Dictionary of Sci-*

entific Biography 9:198-230; (1983). Maxwell's scientific creativity. In R. Aris et al. (note 452), pp. 71-141; M. J. Klein (1970). Maxwell, his demon, and the second law of thermodynamics. *Amer. Sci.* 58:84-97; M. Goldman (1982). *The Demon in the Aether: The Story of James Clerk Maxwell.* Edinburgh: Harris.

23. Quoted from L. P. Williams (1965) (note Ch.9:1), pp. 510-511.

24. J. C. Maxwell (1873). *A Treatise on Electricity and Magnetism.* 2 vols, vol. 1, p. i. Oxford University Press.

25. H. Helmholtz (1928). On the modern development of Faraday's conception of electricity. In *Faraday Memorial Lectures.* pp. 132-159 (145). London: Chemical Society.

26. Ibid., pp. 158-159.

27. G. J. Stoney (1881). On the physical units of nature. *Phil. Mag.* [5] 11:381-390. See J. G. O'Hara (1975). George Johnstone Stoney, F.R.S., and the concept of the electron. *Notes Roy. Soc.* 29:265-276; M. H. Towns and D. A. Davenport (1991). From electrochemical equivalent to a mole of electrons. *Bull. Hist. Chem.* 11:92-101.

28. Rayleigh (1943). *The Life of Sir J. J. Thomson.* Cambridge University Press; J. L. Heilbron (1976). Joseph John Thomson. *Dictionary of Scientific Biography.* 13:362-372.

29. J. J. Thomson (1883), *A Treatise on the Motion of Vortex Rings.* London: See R. H. Silliman (1963). William Thomson: smoke rings and nineteenth-century atomism. *Isis* 54:461-474; S. B. Sinclair (1987). J. J. Thomson and the chemical atom: From ether vortex to atomic decay. *Ambix* 34:89-116; S. Feffer (1989). Arthur Schuster, J. J. Thomson and the discovery of the electron. *Hist. Stud. Phys. Sci.* 20:33-61.

30. A. B. Pippard (1997). J. J. Thomson and the discovery of the electron. In *Electron: A Centenary Volume* (M. Springford, ed.), p. 1. Cambridge University Press. See D. W. Kim (1995). J. J. Thomson and the emergence of the Cavendish school. *Brit. J. Hist. Sci.* 28:191-226.

31. E. N. Hiebert (1995). Electric discharge in rarified gases: The dominion of experiment, Faraday, Plücker, Hittorf. In *No Truth Except in the Details* (A. J. Kox and D. Siegel, eds), pp. 95-134. Dordrecht: Reidel; N. Robotti (1997). The discovery of the electron: I. *Eur. J. Physics* 18:133-138; P. F. Dahl (1997). *Flash of the Cathode Rays* Institute of Physics: Bristol.

32. E. Rutherford and J. J. Thomson (1896). On the passage of electricity through gases exposed to Röntgen rays. *Phil. Mag.* [5] 42:392-407.

33. J. J. Thomson (1897). Cathode rays. *Phil. Mag.* [5] 44:293-310. See I. Falconer (1987). Corpuscles, electrons and cathode rays: J. J. Thomson and the "discovery of the electron." *Brit. J. Hist. Sci.* 20:241-276; T. Aribatzis (1996). Rethinking the discovery of the electron. *Stud. Hist. Phil. Mod. Phys.* 27:405-436.

34. E. Weichert (1897). Über das Wesen der Elektricität. *Schr. Physil.-Ökon. Ges. Königsberg* 38:3-16 (3). See E. A. Davis and I. J. Falconer (1997). *J. J. Thomson and the Discovery of the Electron.* London: Taylor & Francis; O. Darrigol (1998). Aux confins de l'électro-dynamique maxwellienne: Ions et électrons vers 1897. *Rev. Hist. Sci.* 51:5-34.

35. P. Zeeman (1897). On the influence of magnetism on the nature of light emitted by a substance. *Phil. Mag.* [5] 43:226-239. See A. J. Kox (1997). The discovery of the electron: II. The Zeeman effect. *Eur. J. Physics* 18:139-144.

36. J. Z. Buchwald (1985). *From Maxwell to Microphysics.* University of Chicago Press;. O. Darrigol (1994). The electron theories of Larmor and Lorentz. *Stud. Hist. Phil. Sci.* 24:265-336.

37. L. A. DuBridge and P. S. Epstein (1959). Robert Andrews Millikan. *Biog. Mem. Natl. Acad. Sci.* 33:241-282.

38. J. J. Thomson (1899). On the masses of ions in gases at low pressure. *Phil. Mag.* [5] 48:547-567 (547).

39. J. J. Thomson (1907). On rays of positive electricity. *Phil. Mag.* [6] 13:561-575.

40. E. Rutherford (1911). The scattering of α and β particles by matter and the structure of the atom. *Phil. Mag.* [6] 21:669-688.

41. T. J. Trent (1975). Frederick Soddy. *Dictionary of Scientific Biography.* 12:504-509; W. H. Brock (1970). Francis William Aston. Ibid., 1:320-322. M. Cohen (1997). Ernest Rutherford and Frederick Soddy: A historic partnership. *Chem. Intell.* 3 (2):33-40.

42. J. J. Thomson (1904). *Electricity and Matter.* New Haven: Yale University Press; The structure of the atom. *Phil. Mag,* [6] 7:237-265; (1907). *The Corpuscular Theory of Matter.* New York: Scribner's.

43. J. J. Thomson (1914). The forces between atoms and chemical affinity. *Phil. Mag.* [6] 27:757-789; (1923). *The Electron in Chemistry.* Philadelphia: Franklin Institute.

44. C. A. Russell (1971). *The History of Valency.* Leicester University Press; A. N. Stranges (1982). *Electrons and Valence.* College Station, TX: Texas A & M Press; M. Saltzman (1973). J. J. Thomson and the modern revival of dualism. *J. Chem. Educ.* 50:59-61; F. Kober

(1982). Die Theorie der chemischen Bindung in "vorquantenmechanischer" Zeit. *Chem. Z.* 106:1-11; T. Arabatzis and T. Gavroglu (1997) The chemists' electron. *Eur. J. Physics* 18:150-165.

45. W. Nernst (1897). Die elektrolytische Zersetzung wässeriger Lösungen *Ber. chem. Ges.* 30:1547-1563.

46. W. Abegg and G. Bodländer (1899). Die Elektroaffinität, ein neues Prinzip der chemischen Systematik. *Z. anorg. Chem.* 20:453-499; W. Abegg (1904). Die Valenz und das periodisches System: Versuch einer Theorie der Molekülverbindungen. Ibid., 39:330-380 (343).

47. K. G. Falk and J. M. Nelson (1910). The electron conception of valence. *J. Am. Chem. Soc.* 32:1637-1654; H. S. Fry (1915). The electronic conception of positive and negative valences. Ibid., 37:2368-2373; W. A. Noyes (1918). The electron theory. *J. Franklin Inst.* 185:59-84; J. Stieglitz (1922). The electron theory of valence as applied to organic compounds. *J. Am. Chem. Soc.* 44:1293-1313, 1833-1834.

48. G. N. Lewis (1916). The atom and the molecule. *J. Am. Chem. Soc.* 38:762-785; (1923). *Valence and the Structure of Atoms and Molecules.* New York: Chemical Catalog Co. See R. E. Kohler, Jr. (1971). The origin of G. N. Lewis's theory of the shared pair bond. *Hist. Stud. Phys. Sci.* 3:343-376.

49. J. H. Hildebrand (1958). Gilbert Newton Richards. *Biog. Mem. Natl. Acad. Sci.* 31:210-235; J. W. Servos (1984). G. N. Lewis: The disciplinary setting. *J. Chem. Educ.* 61:5-10; M. Calvin and G. T. Seaborg (1984). The College of Chemistry in the G. N. Lewis era: 1912-1948. Ibid., 61:11-13; G. T. Seaborg (1995). Gilbert Newton Lewis—Some personal recollections of a chemical giant. *Chem. Intell.* 1 (3):27-37.

50. G. N. Lewis and M. Randall (1923). *Free Energy and the Thermodynamics of Chemical Substances.* New York: McGraw-Hill. See J. H. Wolfenden (1972). The anomaly of strong electrolytes. *Ambix* 19:175-196.

51. M. Calvin (1984). Gilbert Newton Lewis. His influence on physical-organic chemists at Berkeley. *J. Chem. Ed.* 61:14-18; M. Kasha (1984). The triplet state. An example of G. N. Lewis' research style. Ibid., 61:204-215.

52. Letter from Lewis to Langmuir, 9 July 1919; Lewis Archive, Berkeley. Quoted from A. N. Stranges (1984). Reflections on the electron theory of the chemical bond. *J. Chem. Educ.* 61:185-190 (1984). See R. E. Kohler, Jr. (1972). Irving Langmuir and the octet theory of valence. *Hist. Stud. Phys. Sci.* 4:39-87; (1975). The Lewis-Langmuir

theory of valence and the chemical community. Ibid., 6:431-468; W. B. Jensen (1984). Abegg, Lewis, Langmuir and the octet rule. *J. Chem. Educ.* 61:191-200.

53. I. Langmuir (1939). The structure of proteins. *Proc. Physical Soc.* 51:592-612. See L. Pauling and C. Niemann (1939). The structure of proteins. *J. Am. Chem. Soc.* 61:1860-1867.

54. E. Crawford et al. (1987). *The Nobel Population 1901-1937.* Berkeley: Office for History of Science and Technology.

55. W. Kossel (1916). Über Molekülbildung als Frage des Atombaus. *Ann. Physik* 49:229-362. See H. Kragh (1977). Chemical aspects of Bohr's 1913 theory. *J. Chem. Ed.* 54:208-210.

56. G. N. Lewis (1917). The static atom. *Science* 46:297-302 (301-302).

57. W. M. Latimer and W. H. Rodebush (1920). Polarity and ionization from the standpoint of the Lewis theory of valence. *J. Am. Chem. Soc.* 42:1430-1432; D. Quane (1990), The reception of hydrogen bonding by the chemical community. *Bull. Hist. Chem.* 7:3-13.

58. M. Perutz (1992). The significance of the hydrogen bond in physiology. In *The Chemical Bond* (A. Zewail, ed.), pp. 17-30. San Diego: Academic Press; M. J. Nye (2000). Physical and Biologiccal modes of thought in the chemistry of Linus Pauling. *Stud. Hist. Phil. Mod. Physics* 31:475-491.

59. W. Heitler and F. London (1927). Wechselwirkung neutraler Atome and homopolare Bindung nach der Quantenmechanik. *Z. Physik* 44:455-472.

60. Quotation taken from K. J. Laidler (1993). *The World of Physical Chemistry.* p. 343. Oxford University Press.

61. L. Pauling (1928). The application of quantum mechanics to the structure of the hydrogen molecule and the hydrogen molecule-ion and related problems. *Chem. Revs.* 5:173-213; (1931). The nature of the chemical bond. I. Application of results obtained from the quantum mechanics and from a theory of paramagnetic susceptibility to the structure of molecules. *J. Am. Chem. Soc.* 53:1366-1400. See G. V. Bykov (1965). Historical sketch of the electronic theories of organic chemistry. *Chymia* (1990). Valence-bond theory and chemical structure. *J. Chem. Educ.* 67:633-637; K. Gavroglu and A. Simoes (1994). The Americans, the Germans, and the beginnings of quantum chemistry. *Hist. Stud. Phys. Biol. Sci.* 25:47-110; J. D. Dunitz (1996). Linus

Carl Pauling. *Biog. Mem. Fell. Roy. Soc.* 42:317-338; B. S. Park (1999). Chemical translators: Pauling, Wheland and their strategies for teaching the theory of resonance. *Brit. J. Hist. Sci.* 32:23-46.

62. R. S. Mulliken (1931). Bonding power of electrons and theory of valence. *Chem. Revs.* 9:347-388; J. E. Lennard-Jones (1929). The electronic structure of some diatomic molecules. *Trans. Faraday Soc.* 25:668-686; F. Hund (1977). Early history of the quantum mechanical treatment of the chemical bond. *Angew. Chem. (Int. Ed.)* 16:87-91.

63. E. Hückel (1937). Grundzüge der Theorie ungesättigter und aromatischer Verbindungen. *Z. Elektrochem.* 43:752-788. See H. Hartmann and H. C. Longuet-Higgins (1982). Erich Hückel. *Biog, Mem. Fell. Roy. Soc.* 28:153-162. J. A. Berson (1996). Erich Hückel, pioneer of organic quantum chemistry: Reflections on theory and experiment. *Angew. Chem. (Int. Ed.)* 35:2750-2764; (1999). *Chemical Creativity* pp. 18-24, 33-75. Weinheim: Wiley-VCH; S. Kikuchi (1997). A history of the structural history of benzene—The aromatic sextet rule and Hückel's rule. *J. Chem. Educ.* 74:194-201; S. G. Brush (1999). Dynamics of theory change in chemistry. Part I. The benzene problem 1865-1945. *Stud. Hist. Phil. Sci.* 30:21-81; Part 2. Benzene and molecular orbitals, 1945-1980. Ibid., 30:263-302. See also F. Straus (1927). Johannes Thiele. *Ber. chem. Ges.* 60A:75-13

64. C. Coulson (1952). *Valence*. Oxford: Clarendon Press (3rd ed., 1979); M.J.S. Dewar (1992). *A Semi-empirical Life*. Washington, D.C.: American Chemical Society. A. Simoes and K. Gavroglu (1999). Quantum chemistry *qua* applied mathematics. The contributions of Charles Alfred Coulson (1910-1974). *Hist. Stud. Phys. Biol. Sci.* 29:363-406; (2000) Quantum chemistry in Great Britian: Developing a mathematical framework for quantum chemistry. *Stud. Hist. Phil. Mod. Physics* 31:511-548.

65. R. Hoffmann and R. B. Woodward (1968). The conservation of orbital symmetry. *Acc. Chem. Res.* 1:17-22; R.B. Woodward and R. Hoffmann (1970). *The Conservation of Orbital Symmetry*. New York: Academic Press.

66. G. Herzberg (1985). Molecular spectroscopy. A personal history. *Ann. Rev. Phys. Chem.* 36:1-10.

67. I. I. Rabi (1942). Streams of atoms. *Sci. Am.* 30:265-274; N. F. Ramsay (1953). *Nuclear Moments*. New York: Wiley; S. G. Schlichter (1998). The golden anniversary of nuclear magnetic resonance: NMR—fifty years of surprises. *Proc. Am. Phil. Soc.* 142:533-556.

68. Y. M. Rabkin (1987). Technological innovation in science. The adoption of infrared spectroscopy by chemists. *Isis* 78:31-54.

69. W. L. Bragg (1975). *The Development of X-ray Analysis* (D. C. Phillips and H. F. Lipson, eds.). New York: Hafner; J. M. Robertson (1972). Molecules and crystals. *Helv. Chim. Acta* 55:119-127; G. Ferry (1998). *Dorothy Hodgkin: A Life.* London: Granta.

70. L. S. Ettre and A. Zlatkis (eds.) (1979). *75 Years of Chromatography—A Historical Dialogue.* Amsterdam: Elsevier; A. O. Nier (1955). Determination of isotope masses and abundances by mass spectrometry. *Science* 121:737-744; A. T. Krebs (1955). Early history of the scintillation counter. *Science* 122:17-18.

71. K. J. Laidler and M. C. King (1983). The development of transition-state theory. *J. Phys. Chem.* 87:2657-2664; P. R. Brooks (1988). Spectroscopy of transition region species. *Chem. Revs.* 88:407-428.

72. J. Wislicenus (1887) (note Ch.7:40). A. Michael (1887). Über die Addition von Natriumacetessig- und Natriummalonsäureäthern zu den Äthern ungesättigter Säuren. *J. Prakt. Chem.* 35:349-356. See L. F. Fieser (1975). Arthur Michael. *Biog. Mem. Nat. Acad. Sci.* 46:331-366.

73. P. J. Ramberg (1995). Arthur Michael's critique of stereochemistry, 1887-1899. *Hist. Stud. Phys. Biol. Sci.* 26:89-138.

74. A. B. Costa (1971). Arthur Michael (1853-1942). The meeting of thermodynamics and organic chemistry. *J. Chem. Ed.* 48:243-246.

75. T. B. Johnson (1926). Organic chemistry. In *A Half-Century of Chemistry in America.* (C. A. Browne, ed.), 129-151 (137).

76. J. D. Roberts (1996). The beginnings of physical organic chemistry in the United States. *Chem. Intell.* 2 (2):29-38.

77. M. Wolfrom (1960). John Ulric Nef. *Biog. Mem. Natl. Acad. Sci.* 34:204-227; W. A. Noyes (1941). Julius Stieglitz. Ibid., 21:275-314.

78. See M. J. Nye (1996). "Plus commode et plus elegant": The Paris school of organic reactions in the 1920s and 1930s. *Bull. Hist. Chem.* 19:58-65.

79. E.C.C. Baly (1943). John Norman Collie. *Biog. Mem. Fell. Roy. Soc.* 4:329-356.

80. F. Arndt et al. (1924). Über Dipyrilene und über die Bildungsverhältnisse im Pyron-Ringsystem. *Ber. chem. Ges.* 57:1903-1911. See W. Walter and B. Eistert (1975). Fritz Arndt. *Chem. Ber.* 108:i-xliv; B. Campaigne (1959). The contributions of Fritz Arndt to resonance theory. *J. Chem. Educ.* 36:336-339.

81. N. Bjerrum (1923). Die Konstitution der Ampholyte, besonders

der Aminosäuren, und ihre Dissoziationskonstanten. *Z. physik. Chem.* 104:147-173. See G. B. Kauffman (1980). Niels Bjerrum (1879-1958): a centennial evaluation. *J. Chem. Educ.* 57:863-867.

82. A. Lapworth (1898). A possible basis of generalization of intramolecular changes in organic compounds. *J. Chem Soc. Trans* 73:445-459. (1901). The form of change in organic compounds, and the function of α-meta-orientating groups. Ibid., 79:1265-1284. See M. D. Saltzman (1972). Arthur Lapworth: The genesis of reaction mechanism. *J. Chem. Educ.* 49:750-752.

83. R. Robinson (1947). Arthur Lapworth. *Obit. Not. Fell. Roy. Soc.* 5:555-572; K. Schofield (1995). Some aspects of the work of Arthur Lapworth. *Ambix* 42:160-186.

84. A. Lapworth (1920). Latent polarities of atoms and mechanism of reaction, with special reference to carbonyl compounds. *Mem. Manchester Lit. Phil. Soc.* 64 (3):1-16 (3).

85. C. B. Allsop and W. A. Waters (1947). Thomas Martin Lowry. In *British Chemists* (A. Findlay and W. H. Mills, eds.), pp. 402-428. London: Chemical Society.

86. See J. A. Berson (1999) (note Ch.9:63), pp. 114-115.

87. T. M. Lowry (1923). Intramolecular ionization in organic compounds. *Trans. Faraday Soc.* 19:487-496; J. N. Brønsted (1923). Einige Bemerkungen über den Begriff der Säuren und Basen. *Rec. Trav. Chim. Pays-Bas* 42:718-728. See R. P. Bell (1941). *Acid-Base Catalysis.* Oxford: Clarendon Press.

88. M. D. Saltzman (1997). Thomas Martin Lowry and the mixed multiple bond. *Bull. Hist. Chem.* 20:10-17.

89. T. M. Lowry (1923). Studies of Electrovalency. Part I. The polarity of double bonds. *J. Chem. Soc.* 123:822-831.

90. T. M. Lowry (1935). *Optical Rotatory Power.* London: Longmans, Green.

91. J. Shorter (1998). Some pioneers of the kinetics and mechanism of organic reactions. *Chem. Soc. Revs.* 27:355-366.

92. See W. A. T[ilden] (1911). Nikolai Aleksandrovich Menshutkin. *J. Chem. Soc.* 99:1660-1666.

93. K. Dimroth (1967). Hans Meerwein. *Chem. Ber.* 100:lv-xciv.

94. R. Criegee (1966). Hans Meerwein's scientific work. *Angew. Chem. (Int. Ed.).* 5:333-338; J. A. Berson (1999) (note Ch.9:60), pp. 117-136. L. Birladeanu (2000). The story of the Wagner-Moerwein rearrangement. *J. Chem. Ed.* 77:858-863.

95. C. Meinel (1978). *Die Chemie an der Universität Marburg seit Beginn des 19. Jahrhundert*, pp. 371-405. Marburg: Elwert.

96. J. S. Fruton (1950). Ethylene imine. In *Heterocyclic Compounds* (R. C. Elderfield, ed). pp. 61-72. New York: Wiley.

97. K. T. Mysels (1986). René Marcelin. *J. Chem. Educ.* 63:740.

98. K. J. Laidler and M. C. King (1983) (note Ch.9:71); K. J. Laidler (1993) (note Ch.9:60), pp. 243-249; (1998). A lifetime of transition-state theory. *Chem. Intell.* 4 (3):39-47.

99. A. R. Todd and J. W. Cornforth (1976). Robert Robinson. *Biog. Mem. Fell. Roy. Soc.* 22:415-527; R. Robinson (1976). *Memoirs of a Minor Prophet: Seventy Years of Organic Chemistry*. London: Elsevier; T. I. Williams (1990). *Robert Robinson: Chemist Extraordinary*. Oxford University Press.

100. W. O. Kermack and R. Robinson (1922). Explanation of the property of induced polarity of atoms and an interpretation of the theory of partial valencies on an electronic basis. *J. Chem. Soc.* 121:427-440; R. Robinson (1925). Polarization of nitrosobenzene. *Chem. Ind.* 44:456-458; (1947). The development of electrochemical theories of the course of reactions of carbon compounds. *J. Chem. Soc.*, pp. 1288-1302.

101. A. R. Todd and J. W. Cornforth (1976) (note Ch.9:99), p. 467. See S. G. Brush (1999) (note Ch.9:63), pp. 52-53 for a correction of this statement.

102. R. Robinson (1917). A synthesis of tropinone. *J. Chem. Soc.* 111:762-768. See A. J. Birch (1993). Investigating a scientific legend: The tropinone synthesis of Sir Robert Robinson, F.R.S. *Notes Roy. Soc.* 47:277-296.

103. R. Robinson (1917). A theory of the mechanism of the phyto-chemical synthesis of certain alkaloids. *J. Chem. Soc.* 111:876-899. See E. Leete (1965). Biosynthesis of alkaloids. *Science* 147:1000-1006; A. J. Birch (1990). Chance and design in biosynthesis. *Interdisc. Sci. Revs.* 1:215-233. In 1948, I asked Robinson whether he was awaiting with interest the outcome of radioisotope experiments to test the validity of his biosynthetic hypotheses. His reply, as nearly as I can recall, was, "Of course they are correct." My presumption did no harm, for during our next stay in Oxford my wife and I were invited to tea at the Robinson's, and taken on a conducted plant-by-plant tour of their elegant garden.

104. R. Robinson (1934). Structure of cholesterol. *Chem. Ind.* pp. 1062-1063; (1955). *The Structural Relations of Natural Products*. Oxford: Clarendon Press.

105. L. Ruzicka (1956). Bedeutung der theoretischen organischen Chemie für die Chemie der Terpenverbindungen. In *Perspectives in Organic Chemistry* (A. Todd, ed.), pp. 265-314. New York: Interscience. See V. Prelog and O. Jeger (1980). Leopold Ruzicka. *Biog. Mem. Fell. Roy. Soc.* 26:411-501.

106. V. Prelog (1950). Newer developments of the chemistry of many-membered ring compounds. *J. Chem. Soc.*, pp. 420-428; (1956). Bedeutung der vielgliedrigen Ringverbindungen. In *Perspectives in Organic Chemistry*, pp. 96-133. New York: Interscience; D.H.R. Barton (1950). The conformation of the steroid nucleus. *Experientia* 6:316-320; (1991). *Some Recollections of Gap Jumping.* Washington, D.C.: American Chemical Society. See also O. Bastiansen (1982). Odd Hassel. *Chem. Brit.* 18:442.

107. J. Shorter (1980). A. G. Vernon-Harcourt: A founder of chemical kinetics and friend of "Lewis Carroll." *J. Chem. Educ.* 57:411-416; H. K[ing] (1931). Kennedy Joseph Previté Orton. *J. Chem. Soc.* pp. 1042-1048; J.F.J. Dippy (1976). Herbert Ben Watson. *Chem. Brit.* 12:227-228.

108. C. K. Ingold (1964). Edward David Hughes. *Biog. Mem. Fell. Roy. Soc.* 10:147-182.

109. C. W. Shoppee (1972). Christopher Kelk Ingold. *Biog. Mem. Fell. Roy. Soc.* 18:349-411; K. T. Leffek (1996). *Sir Christopher Ingold: A Major Prophet of Organic Chemistry.* Victoria, B.C.: Nova Lion Press.

110. B. Flürscheim (1909). The relation between the strength of acids and bases and the quantitative distribution of affinity in the molecule. *J. Chem. Soc. Trans.* 95:718-734 (718). See C. K. Ingold (1956). Bernard Flürscheim. *J. Chem. Soc.* pp. 1087-1089.

111. M. D. Saltzman (1980). The Robinson-Ingold controversy. *J. Chem. Educ.* 57:484-488; K. Schofield (1994). The development of Ingold's system of organic chemistry. *Ambix* 41:87-107; J. F. Bunnet (1996). Physical organic terminology, after Ingold. *Bull. Hist. Chem.* 19:33-42.

112. C. K. Ingold (1934). Principles of an electronic theory of organic reactions. *Chem. Revs.* 15:225-274.

113. Quoted from C. W. Shoppee (1972) (note Ch.9:109) pp. 356-357. This statement also appears in the article by D.H.R. Barton (1996). Ingold, Robinson, Winstein, Woodward, and I. *Bull. Hist. Chem.* 19:43-47 (44). The source is incorrectly given as Ingold's 1934 review (note Ch.9:109).

114. C. K. Ingold (1953). *Structure and Mechanism in Organic Chemistry.* Ithaca: Cornell University Press. 2nd ed., 1969.

115. A. Todd and J. W. Cornforth (1976) (note Ch.9:99), p. 466.

116. E. Fischer (1923). *Untersuchungen über Aminosäuren, Polypeptide, und Proteine. II (1907-1919).* pp. 736-892 (775). Berlin: Springer.

117. See E. E. Turner (1962). Joseph Kenyon. *Biog. Mem. Fell. Roy. Soc.* 8:49-66; J. A. Berson (1999) (note Ch.9:63), pp. 150-160.

118. E. D. Hughes et al. (1935). Aliphatic substitution and the Walden inversion. Part I. *J. Chem. Soc.* pp. 1525-1530; C. A. Bunton (1966). Nucleophilic substitution and the Walden inversion. In *Studies on Chemical Structure and Reactivity* (J. H. Ridd ed.), pp. 73-102. New York: Wiley.

119. Quoted from R.O.C. Norman and J. H. Jones (1986). William Alexander Waters. *Biog. Mem. Fell. Roy. Soc.* 32:599-627 (609). See J. M. McBride (1974). The hexaphenylethane riddle. *Tetrahedron* 30:2009-2022; W. A. Waters (1984). Some comments on the development of free radical chemistry. *Notes Roy. Soc.* 39:105-124.

120. M. D. Saltzman (1986). The development of physical organic chemistry in the United States and the United Kingdom: 1919-1939, parallels and contrasts. *J. Chem. Educ.* 63:588-593; L. Gortler (1987). The development of a scientific community: Physical chemistry in the United States, 1925-1950. In J. G. Traynham (note Ch.6:53), pp. 95-113; J. D. Roberts (1996). The beginnings of physical organic chemistry in the United States. *Bull. Hist. Chem.* 19:48-56.

121. J. B. Conant (1970). *My Several Lives, Memoirs of a Social Inventor.* p. 33. New York: Harper & Row; (1952). Elmer Peter Kohler. *Biog. Mem. Natl. Acad. Sci.* 27:265-291.

122. J. B. Conant (1932). *Equilibria and Rates of some Organic Reactions,* pp. 26-27. New York: Columbia University Press.

123. G. B. Kistiakowsky and F. H. Westheimer (1979). James Bryant Conant. *Biog. Mem. Fell. Roy. Soc.* 25:209-232; P. D. Bartlett (1983). James Bryant Conant. *Biog. Mem. Natl. Acad. Sci.* 54:91-124; J. G. Hershberg (1993). *James B. Conant: From Harvard to Hiroshima and the Making of the Nuclear Age.* New York: Knopf.

124. J. B. Conant (ed.) (1950). *Robert Boyle's Experiments in Pneumatics,* pp. 5-6. Cambridge, Mass.: Harvard University Press. See also J. B. Conant (1947). *On Understanding Science.* New Haven: Yale University Press.

125. F. Westheimer (1997). Louis Plack Hammett. *Biog. Mem. Natl. Acad. Sci.* 72:137-149.

126. T. S. Moore (1936). The Hantzsch memorial lecture. *J. Chem. Soc.* pp. 1051-1066.

127. L. P. Hammett (1966). Physical organic chemistry in retrospect. *J. Chem. Educ.* 43:464-469 (468).

128. L. P. Hammett and A. J. Deyrup (1932). A series of simple basic indicators. I. The acidity functions of mixtures of sulfuric acids with water. *J. Am. Chem. Soc.* 54:2721-2739. L. P. Hammett (1935). Reaction rates and indicator acidities. *Chem. Revs.* 16:67-79.

129. L. P. Hammett (1935). Some relations between reaction rates and equilibrium constants. *Chem. Revs.* 17:125-136; J. N. Brønsted and K. Pedersen (1924). Die katalytische Zersetzung des Nitramids und ihre physikalisch-chemische Bedeutung. *Z. physik. Chem.* 108:185-235.

130. H. H. Jaffé (1953). A reexamination of the Hammett equation. *Chem. Revs.* 53:191-261; J. Shorter (1996). The sigma culture. *Chem. Intell.* 2 (1):39-47.

131. L. P. Hammett (1940). *Physical Organic Chemistry.* New York: McGraw Hill. (2nd ed., 1970).

132. J. C. Bailar, Jr. (1970). Moses Gomberg. *Biog. Mem. Natl. Acad. Sci.* 41:141-173. A. J. Ihde et al. (1966), International symposium on free radicals. *Chem. & Eng. News.* Oct. 3, pp. 90-107. See McBride (1974) (note Ch.9:119).

133. J. D. Roberts (1974). James Flack Norris. *Biog. Mem. Natl. Acad. Sci.* 45:413-426.

134. F. H. Westheimer (1960). Morris Selig Kharasch. *Biog. Mem. Natl. Acad. Sci.* 34: 123-152.

135. W. G. Young and S. Winstein (1973). Howard Johnson Lucas. *Biog. Mem. Natl. Acad. Sci.* 34:163-178.

136. F. H. Westheimer (1998). Paul Doughty Bartlett. *Proc. Am. Phil. Soc.*142:447-456; J. D. Roberts (1998). Paul D. Bartlett. *Chem. Intell.* 4 (2):34-39.

137. P. Bartlett and L. H. Knox (1939). Bicyclic structures prohibiting the Walden inversion. Replacement reactions in 1-substituted 1-apocamphanes. *J. Am. Chem. Soc.* 61:3184-3192.

138. At that time, I was working on the same problem, and had occasion to admire Bartlett's reports; see J. S. Fruton (1950) (note Ch.9:96).

139. J. D. Roberts (1998) (note Ch.9:136), p. 38. See J. D. Roberts (1990). *The Right Place at the Right Time.* Washington, D.C.: American Chemical Society.

140. P. D. Bartlett (1972). The scientific work of Saul Winstein. *J.*

Am. Chem. Soc. 94:2161-2170; W. J. Young and D. J. Cram (1973). Saul Winstein. *Biog. Mem. Natl. Acad. Sci.* 34:321-353.

141. R. Adams and S. Winstein (1948). The role of neighboring groups in replacement reactions. XIV. The 5,6-double bond in cholesteryl *p*-toluene sulfonate as a neighboring group. *J. Am. Chem Soc.* 70:838-840.

142. S. Winstein (1951). Neighboring groups in displacement and rearrangement. *Bull. Soc. Chim.* C:55-61; P. D. Bartlett (1965). *Nonclassical Ions: Reprints and Commentary.* New York: Benjamin; A. Winstein (1969). Nonclassical ions and homoaromaticity. *Quart. Revs. Chem. Soc.* 23:141-176.

143. S. Winstein (1965). On Brown's classical norbornyl cation. *J. Am. Chem. Soc.* 87:381-382.

144. For example, see L. de Vries and S. Winstein (1960). Neighboring carbon and hydrogen. XXXIX. Complex rearrangements of bridged ions. Rearrangement leading to the bird-cage hydrocarbon. *J. Am. Chem. Soc.* 82:5363-5376.

145. A. Todd and J. Cornforth (1981). Robert Burns Woodward. *Biog. Mem. Fell. Roy. Soc.* 27:629-695 (there is no memoir for Woodward in the Biographical Memoirs of the National Academy of Sciences); M. E. Bowden and T. Benfey (1992). *Robert Burns Woodward and the Art of Organic Synthesis.* Philadelphia: Beckman Center for the History of Chemistry.

146. R. B. Woodward (1956). In A. Todd (ed.) (note Ch.9:103), pp. 155-184 (155).

147. At the end of one of his public lectures, Woodward said: "It remains to thank with all the warmth at my command those who fought and enjoyed the battle with me." See W. D. Ollis (1980). Robert Burns Woodward—an appreciation. *Chem. Brit.* 16:210-216 (214).

148. C. E. Woodward (1989). Art and elegance in the synthesis of organic compounds: Robert Burns Woodward. In *Creative People at Work* (D. B. Wallace and H. E. Gruber, eds.), pp. 227-252. New York: Oxford University Press.

149. R. B. Woodward (1966). Recent advances in the chemistry of natural products. *Science* 153:487-493. See L. B. Slater (2000). Industry and academy: The synthesis of steroids. *Hist. Stud. Phys. Biol. Sci.* 30:443-480. G. Mulheim (2000). Robinson, Woodward, and the synthesis of cholesterol. *Endeavour* 24:107-110.

150. R. B. Woodward (1965). The total synthesis of colchicine. *Harvey Lectures* 59:31-47.

151. G. Stork (1980). R. B. Woodward, 1917-1979. *Nature* 284:383-384.

152. J. A. Berson (1999) (note Ch.9:63), pp. 82-103.

153. R. B. Woodward and K. Bloch (1953). The cyclization of squalene in cholesterol synthesis. *J. Am. Chem. Soc.* 75:2023-2024.

154. R. B. Woodward and R. Hoffmann (1965). Stereochemistry of electrocyclic reactions. *J. Am. Chem. Soc.* 87:395-397. See I. Fleming (1999). *Pericyclic Reactions.* Oxford University Press.

155. R. Hoffmann and R. B. Woodward (1970). Orbital symmetry control of chemical reactions. *Science* 167:825-831. See also note Ch.9:65. I. Hargittai and M. Hargittai (1987) (note Ch.7:47), and E. Heilbronner and J. D. Dunitz (1993) (note Ch.7:47).

156. J. A. Berson (1999) (note Ch.9:63), pp. 24-30; T. Yonezawa (1999). Professor Kenichi Fukui. *Chem. Intell.* 5 (1):38-42.

157. K. C. Nicolaou, D. Vourloumis, N. Winssinger, and P. S. Baran (2000). The art and science of total synthesis at the dawn of the twenty-first century. *Angew. Chem. (Int. Ed.)* 39:49-122.

Conclusion

1. S. de Chadarevian and H. Kamminga (eds.) (1998). *Molecularizing Biology and Medicine: New Practices and Alliances, 1910s-1970s.* Amsterdam: Harwood.

2. Ibid., p. 1.

3. K. C. Nicolaou, E. J. Sorensen, and N. Winssinger (1998). "The art and science of organic and natural products synthesis," *J. Chem Educ.* 75:1226-1258; D. J. Austin (2000). The chemistry and biology interface: an emerging field or an old friend? *Cell* 101:125-126.

INDEX OF
PERSONAL NAMES

GENERAL INDEX